U0344413

本丛书名由中国科学院院士母国光先生题写

光学与光子学丛书

光学与光子学丛书

啁啾脉冲在光纤中的传输

郑宏军　黎　昕　白成林　著

科学出版社

北　京

内 容 简 介

脉冲及其传输的性能研究是激光、光电子及光通信领域的永恒话题，其应用广泛。本书阐述了光纤通信系统中的关键光电子器件基础、不同光纤的传输系统特性，分析了脉冲传输理论基础与数值计算方法，利用数值计算和实验研究了啁啾光脉冲在不同光纤通信系统中的线性和非线性传输特性。本书内容主要包括啁啾脉冲在标准单模光纤系统、线性色散平坦光纤系统、凸形和凹形色散平坦光纤系统、色散离散渐减光纤系统、变参量光纤系统、双折射光纤系统、拉曼放大光纤系统中的传输特性；同时实验研究了模分复用前沿领域中新型少模光纤、新型模分复用器及其模式传输系统的性能。

本书适合激光、光电子、光通信等专业的学生或科技工作者学习和参考。

图书在版编目（CIP）数据

啁啾脉冲在光纤中的传输／郑宏军，黎昕，白成林著. —北京：科学出版社，2018.1

(光学与光子学丛书)

ISBN 978-7-03-056351-4

Ⅰ. ①啁… Ⅱ. ①郑… ②黎…③白… Ⅲ. ①光脉冲-光纤传输技术 Ⅳ. ①TN781②TN818

中国版本图书馆 CIP 数据核字（2018）第 010325 号

责任编辑：钱 俊 胡庆家／责任校对：杜子昂

责任印制：张 伟／封面设计：耕者设计

科学出版社出版

北京东黄城根北街 16 号

邮政编码：100717

http://www.sciencep.com

北京虎彩文化传播有限公司印刷

科学出版社发行 各地新华书店经销

*

2018 年 1 月第 一 版 开本：720×1000 1/16

2019 年 1 月第三次印刷 印张：12 1/4

字数：240 000

定价：89.00 元

（如有印装质量问题，我社负责调换）

前　言

脉冲及其传输的性能研究是光通信研究的永恒话题，但鉴于实验条件等因素的限制，以往的脉冲传输与测量实验大多采用自相关技术等测量脉冲时域特性，不能准确得到脉冲的时域波形等特性。考虑到脉冲通常具有较大的频率啁啾且可以通过预啁啾技术等调节啁啾，提出采用能够准确测量脉冲时域波形等特性的二次谐波频率分辨光学门(SHG-FROG)前沿热点技术实验研究啁啾光脉冲在不同光纤通信系统中线性和非线性传输特性并作了深入系统的理论研究；数值分析与实验研究了啁啾脉冲在标准单模光纤系统、线性色散平坦光纤系统、凸形和凹形色散平坦光纤系统、色散离散渐减光纤系统、变参量光纤系统、双折射光纤系统、拉曼放大光纤系统中的传输特性；实验研究了模式复用前沿领域中新型少模光纤、新型模分复用器及其模式传输系统的性能；为光纤通信系统及其关键光电子器件设计和优化提供了重要的理论和实验依据。

本书共 8 章。第 1 章主要介绍啁啾脉冲传输若干关键技术研究现状、主要问题以及脉冲传输理论基础与数值计算方法；第 2 章深入分析啁啾脉冲实验测量原理与啁啾脉冲自相关特性；第 3 章理论分析与实验研究啁啾脉冲的线性传输特性；第 4 章揭示啁啾脉冲在标准单模光纤通信系统中的非线性传输演化规律；第 5 章在研究啁啾脉冲线性与非线性传输机理基础上，进一步实验和数值研究啁啾脉冲在色散平坦光纤正常色散区、色散离散渐减光纤系统、变参量光纤系统、双折射光纤系统等特种光纤系统中的传输性能，重点分析了脉冲输入功率、脉冲啁啾参量、脉冲间隔、光纤系统色散参量、非线性参量等对传输特性的影响；第 6 章研究啁啾脉冲在凸形和凹形色散平坦光纤系统中所产生超连续谱的特性；第 7 章讨论拉曼放大对啁啾脉冲传输特性的影响，提出了利用啁啾脉冲传输实验测量拉曼增益放大系数的新方法；第 8 章提出并实验研究模式复用前沿领域中新型少模光纤、新型模分复用器及其模式传输系统的性能。

本书相关工作得到了上海交通大学胡卫生教授及其课题组、University of Central Florida (CREOL) 的 Guifang Li 教授及其课题组、北京交通大学吴重庆教授及其课题组、华中科技大学徐静平教授及其课题组、刘山亮教授的热情指导和帮助！在此表示衷心感谢！同时本书相关工作得到了国家自然科学基金(项目编号：61671227，61431009，60778017 和 60877057)，山东省自然科学基金(项目编号：ZR2011FM015)，“泰山学者”建设工程专项经费，山东省教育厅重点科技计划项目(项目编号：J05C09)，“区域光纤通信网与新型光通信系统国家重点实验室”开放基金(项目编号：2011GZKF031101)等的资助，在此一并表示衷心感谢！

本书是我们研究工作的初步梳理和总结，由于水平有限，再加上光纤通信技术及其应用迅猛发展、日新月异，书中难免存在不妥之处，敬请各位读者批评指正。

作　者

2017年12月

目　录

第1章　绪　论

自 1960 年美国科学家 Maiman 成功研制世界上第一台激光器(红宝石激光器)[1]以来，固体、液体、气体和半导体等各种激光器在短时间内相继问世，为科学研究提供了崭新的有效工具。技术上的重大突破，催生了光电子学和激光物理学等系列新学科。之后，激光及其光电子技术取得了迅速的发展，并广泛应用于工农业生产[2-4]、生命科学[5]、信息科学[6-13]、强场物理[14-24]、“快点火”受控核聚变[25-32]、探索极端物理条件下物质行为等重大科学研究领域，在整个经济及社会发展中发挥出越来越重要的作用。随着激光及其光电子技术的发展，至 1970 年，人们已经将光纤损耗降低到约 20dB/km；到 1979 年，将 1.55μm 波长的光纤损耗降低到约 0.2dB/km[33]。低损耗光纤的获得，不仅导致了光纤通信领域的革命，而且也导致了非线性光纤光学新领域的出现。随后，掺铒光纤放大器得到了迅速发展[34-48]，有效补偿了 1.55μm 波长区的光纤损耗，有力地推动了线性和非线性光纤光学的快速发展，激光光源(半导体激光光源、光孤子光源和波分复用光纤通信系统的超连续谱光源等)，放大器(半导体光放大器、掺铒光纤放大器和拉曼放大器等)，光电探测器件，调制器、波分复用和解复用器、耦合器和光开关器件等都取得了突破性进展，光纤通信技术不断得到阶段性的飞跃。大多数光电子器件中使用了光纤，与光纤通信系统及其关键光电子器件相关的激光脉冲在光纤中的线性和非线性传输特性研究得到了国际科技工作者的普遍关注[49-149]。

1.1　脉冲传输若干关键技术的国内外研究现状

为了研究脉冲在线性和非线性传输过程中的波形、相位和啁啾(chirp)等特性，人们进行了长期的积极探索。脉冲线性传输的研究在光纤通信等领域中仍然占据重要地位，并不断取得高速率、大容量的新成果[51-78]。例如，文献[51]采用波分复用超连续谱光源、CS-RZ 码流、色散管理技术完成了 81×40Gbit/s、80km 的线性传输实验，误码率达到 10^{-9}。文献[52]采用超连续谱产生的 1046 个信道光源(信道间隔 6.25GHz)在单模光纤中完成了 1046×2.67Gbit/s、126km 的超密集波分复用传输实验。文献[53—55]分别研究了 PAM4 调制格式 2×64Gbit/s 标准单模光纤

70km 传输，无源光网络 PAM4 调制格式突发模式 25Gbit/s 数据上行传输，PAM4 调制格式 4×56.25Gbit/s 标准单模光纤 80km 传输。文献[56—60]分别分析了 16QAM 调制格式、224Gbit/s、单边带直接探测、标准单模光纤 160km 传输，64QAM 调制格式、3.84Tbit/s、单信道奈奎斯特脉冲传输 150km，42.3Tbit/s、18Gbaud、64QAM 调制格式、C 波段、WDM 相干传输 160km，PDM-16QAM 调制格式、65-Gbaud、1.6Tbit/s(4×400Gbit/s)标准单模光纤 205km 数据上行传输，QAM-PAM 调制格式、280Gbit/s、偏振复用传输 320km。文献[61]和[62]研究了少模模分复用传输；文献[63—65]对用于模分复用传输的光子灯笼进行了研究；文献[66]和[67]分别研究了 3×10Gbit/s、20km 少模模式组传输，5Tbit/s、2.2km 少模模式组传输；文献[68]和[69]研究了轨道角动量模式传输；文献[70—78]研究了啁啾脉冲产生与传输放大。

自 1973 年，Hasegawa 在脉冲非线性传输中提出光孤子概念以来[79]，光孤子脉冲的非线性传输理论和实验得到了普遍研究[79-149]。1974 年，文献[85]提出采用逆散射方法求解非线性问题，随后孤子脉冲的非线性理论不断得到完善。1980 年，Mollenauer 等在脉冲传输实验中发现光孤子，极大地促进了脉冲非线性传输的研究[86]。为了将孤子脉冲周期性地放大传输，就必须克服光纤损耗。目前，通常采用的孤子放大方法有掺铒光纤放大器的集总放大方式、分布式拉曼放大方式或前两者相结合的方式。在国内外光孤子通信系统的研究与开发中，采用这些方法已进行了一系列现场实验并取得了可喜成果。文献[87]实现了单信道 5Gbit/s、15000km 和双通道 10Gbit/s、11000km 的孤子传输实验。文献[88]实现了 4×20Gbit/s、2000km[89]和 16×20Gbit/s、1300km 的波分复用孤子传输实验。文献[90]采用色散补偿技术进行了单信道 20Gbit/s、传输距离为 2000km 的光孤子东京城域网传输实验；文献[91]实现了 2×80Gbit/s、10000km 的孤子传输实验。文献[92]采用色散管理孤子技术完成了 40Gbit/s、1000km 的传输实验。Gouveia-Neto, Iwatsuki, Murphy 和 Mollenauer 等利用拉曼放大进行了孤子脉冲产生[93]、啁啾孤子脉冲压缩[94-95]和孤子脉冲传输的研究[96-105]。其中，Murphy 利用拉曼放大将 20ps 啁啾孤子压缩为 1.3ps[95]；Mollenauer 等利用 1460nm CW 拉曼泵浦实现了 10ps、1560nm 孤子脉冲在 10km 单模光纤中的传输实验[96]。Iwatsuki 等利用 1450nm 和 1480nm CW 双向拉曼泵浦实现了 5Gbit/s、1550nm 孤子脉冲在 23km 色散位移光纤中的传输实验，误码率为 2×10^{-10}[97]。Okhrimchuk 等利用 1240nm 拉曼泵浦和 24km 单模光纤环实现了 10Gbit/s、10000km 孤子传输实验[98]。Ereifej 等利用 1455nm 拉曼泵浦、色散位移光纤和单模光纤混合光纤环将 5ps、10GHz 孤子脉冲调制成 10Gbit/s 码流并复用成 40Gbit/s 孤子码流，实现了 40Gbit/s 孤子码流在色散管理孤子系统中无误码传输 5300km 和 7500km[99]。Pincemin 等研究了拉曼放大对 N×20Gbit/s 波分复用系统中色散管理孤子在信道内相互作用和信道间碰撞的影响，表明拉曼放大能有效减少脉冲信道内相互作用和信道间碰撞的影响[100]。Tio 等数值分析了孤子脉冲在拉曼放大器中的传输[101-105]。文献[106—109]研究了光纤激光器中的非线性孤子脉冲现象；文献[110—117]研究了色散波演化成基阶孤子脉冲、孤子光纤振荡器、中红外孤子脉冲产生、呼吸孤子脉冲、飞秒拉曼孤子脉冲、色散管理孤子脉冲、多级孤子脉冲压缩、孤子与非线性傅里叶变换；文献[118—120]研究了孤子脉冲的自频移现象；文献[121—125]研究了明暗孤子相互作用、双孤子脉冲以及多孤子脉冲的特性和相互作用；文献[126—128]研究了少模光纤、多模光纤中的孤子脉冲相互作用和压缩。

近年来，世界各国相继提出了光孤子通信的发展计划。例如，日本星计划项目(Star Project)，旨在采用孤子技术构建全球距离的 Tbit/s 全光网，以满足急剧增长的多媒体、数据等业务需求，使现有的通信网改建升级为下一代通信网基础设施。美国 MIT 林肯实验室主持超快孤子多接入网计划，该计划研究单信道 100Gbit/s 的 TDM 多接入网的网络结构、节点与收发设备等关键技术。荷兰飞利浦光电子研究中心主持的欧洲升级计划目标是在欧洲网已铺设的标准单模光纤上采用光孤子传输，为欧洲的通信干线增大容量。法国电信(CNET)制订的致力于 WDM 孤子传输技术产业化，实现 1Tbit/s、1000km 孤子传输的科技发展计划等，以及最近提出的孤子通信系统 ITU-T 标准建议等都表明孤子通信系统是下一代光纤通信系统的优选方案。

在国家自然科学基金、国家“863”计划和相关部委等的支持下，中国有许多科研院所开展了光孤子通信理论与实验研究，主要研究了光孤子光源、光孤子补偿放大器、孤子传输理论和传输实验，并取得了许多成果。Liu 等通过引入新的因变量变换给出了修正的非线性薛定谔方程在不同条件下的孤子解，并分析了孤子解的特征[129-131]。高以智、许宝西、杨祥林和余建军等采用半导体光孤子光源，利用掺铒光纤放大器对孤子脉冲放大后，在色散位移光纤中进行了长距离传输研究[132-136]；余建军等在不同色散光纤的光纤链中进行了孤子传输的实验[137]；张晓光等以色散补偿光纤作为色散补偿器件成功实现了 10GHz、38km 色散管理孤子的传输实验[138]。曹文华等研究了拉曼放大作用下的孤子脉冲传输等情况[139-141]；其中，曹文华等计算模拟了在色散缓变光纤中利用拉曼泵浦脉冲产生拉曼孤子脉冲的情况[139]。李宏等数值研究了利用调制拉曼泵浦脉冲来控制暗孤子的传输[140]，表明调制拉曼泵浦进行传输控制，不仅可以有效地抑制暗孤子的时间抖动，同时还明显降低了暗孤子间的相互作用。沈廷根等研究发现[141]，在光子晶体光纤的各个线缺陷中掺入拉曼增益介质，能够对孤子脉冲进行拉曼放大。文献[142—149]研究了矢量孤子、耗散孤子、涡旋光孤子、Peregrine 孤子、近红外波段宽带孤子及其传输放大特性。贾东方、刘颂豪、庞小峰、杨祥林和黄景宁等的著作或译作对我国光孤子理论和实验研究作出了较大的贡献[80-84]。国内光孤子研究的技术基础相对较弱，面临的技术难度较大，研究经费不足等，导致研究工作进展缓慢，与国外研究的差距较大。目前华南师范大学、华中科技大学、北京邮电大学、山西大学、天津大学、内蒙古大学、南京邮电学院、聊城大学等部分单位仍然坚持研究。

1.2 啁啾脉冲传输若干关键技术研究中存在的主要问题

由于受到实验条件等各种因素的限制，在以往的激光脉冲传输与测量实验中大多采用自相关技术等测量脉冲时域特性[80, 86, 93-98, 132-138, 150-168]，不能准确得到脉冲的时域波形、相位和啁啾等信息，而仅注重输入、输出脉冲的宽度及其变化等。电光条纹相机[169]虽然能够较准确地测量光强的空间或时间分布，但受到光谱响应特性限制，不适于测量长波长光脉冲，且价格昂贵、结构复杂、操作苛刻。频率分辨光学门(FROG)测量技术[168, 170-191]经过十几年的研究和发展，可以有效抑制背景且具有较高的动态范围，能够准确测量脉冲的时域波形、相位和啁啾等特征信息，并能较全面地测量各种脉冲特性。就作者所知，目前国内外了解和应用这种技术进行脉冲传输研究的科研人员还为数不多。研究发现，脉冲通常具有较大频率啁啾，其频率啁啾的变化对脉冲特性的影响较大[168, 192-208]，且频率啁啾可以通过改变传输长度、采用啁啾光栅技术或预啁啾技术等进行调节[168]。显然，利用能够准确测量脉冲的时域波形等特性的二次谐波频率分辨光学门(SHG-FROG)技术实验研究啁啾脉冲在光纤通信系统中的线性和非线性传输特性，并对脉冲传输作深入系统的理论研究是一项创新性工作，可以为光纤通信系统及其关键光电子器件设计和优化提供重要的理论和实验依据。

线性传输问题，考虑到以往实验条件等各种因素的限制，利用 SHG-FROG 技术实验研究 10GHz 短脉冲在色散平坦光纤中的线性传输特性，并对脉冲传输作深入的理论研究非常重要。在此研究基础上，有必要进一步研究其他脉冲的线性传输特性。自从在非线性传输中提出光孤子概念以来[79]，关于双曲正割脉冲的研究受到了广泛关注，但是其研究工作大多限于非线性传输中[80-84, 86-149, 206-208]。双曲正割脉冲线性传输特性的变化规律很难解析给出，人们对其线性传输规律的认识受到很大限制。以往对指数脉冲的研究仅限于 $T\geqslant 0$ 的情况[209-211]，双边指数脉冲($-\infty<T<\infty$)的传输特性难于解析研究。鉴于此，研究啁啾双曲正割脉冲和啁啾双边指数脉冲的线性传输特性是一项创新的工作，可以填补人们对两脉冲线性传输认识的国际空白。

非线性传输问题，特别是由非线性薛定谔方程所描述的孤子问题出现在现代科学的各个分支中,例如,信息科学中的光纤孤子可以用非线性薛定谔方程描述。由于光孤子具有广阔、明朗的应用前景和易于实验研究等特点，几十年来得到了广泛的研究和发展，为物理学、信息科学、生命科学、等离子体和其他学科中众多的类似问题的解决作出了重大贡献[80-84]。鉴于以往自相关技术实验条件等各种因素的限制[80, 86, 93-98, 132-138, 150-168]，利用 SHG-FROG 技术实验研究啁啾脉冲在光纤中传输前后脉冲宽度、波形、啁啾和时间带宽积等的变化以及脉冲演化形成孤子的规律和特点并作系统的理论研究具有创新性。在此基础上，进一步研究双边指数脉冲($-\infty<T<\infty$)等的非线性传输特性非常重要。这对光孤子光源及其通信系统的设计和优化具有重要指导作用。

近年来，光纤超连续谱广泛应用于波分复用光通信系统、超连续谱激光源等关键光电子器件[80, 51, 52, 212-225]、超短脉冲产生[226-228]、光学相干层析和光谱分析[229-231]等重要领域，其研究得到了广泛关注。人们对在色散位移光纤、色散平坦光纤、色散平坦渐减光纤、光子晶体光纤等多种光纤中产生的超连续谱进行了理论和实验研究，其中凸形色散分布平坦光纤或平坦渐减光纤中产生的超连续谱特性较好[232-238]。以往的文献通常以无啁啾的双曲正割脉冲为入射脉冲研究光纤中的超连续谱。考虑到实际脉冲源通常具有较大频率啁啾，以啁啾高斯脉冲为入射脉冲，研究线性频率啁啾对啁啾高斯脉冲产生超连续谱的影响，并与啁啾双曲正割脉冲产生超连续谱的情况进行了比较，可为我们利用啁啾脉冲获得最佳超连续谱和实现波分复用光纤通信系统超连续谱光源的优化设计提供重要依据。

孤子脉冲碰撞具有科学上的重要性，在光逻辑器件和偏振复用系统中具有重要的实用价值，从而引起了科技工作者的广泛兴趣[80, 239-250]。Menyuk 推导了孤子脉冲在双折射光纤中的非线性耦合方程，并研究了无啁啾孤子脉冲的传输稳定性[80, 244-246]。随后，无啁啾正交偏振孤子碰撞特性得到了普遍关注[247-250]。考虑到脉冲通常具有较大的频率啁啾，啁啾孤子脉冲在双折射光纤中的碰撞特性研究是一项具有特色的工作，对光逻辑器件和偏振复用系统的进一步研究具有重要意义。

光纤拉曼放大是基于受激拉曼散射原理，以光纤作为增益介质而实现的全光放大，是光纤光学中的一个重要的非线性过程。相对于稀土掺杂的光纤放大而言，它具有更大的增益带宽、灵活的增益谱区、更低的放大器自发辐射噪声以及能够有效抑制信噪比(SNR)的劣化等优点，近年来在光纤传输系统中获得越来越多的应用[80, 93-105, 138-141]，如 Mollenauer 等利用拉曼放大进行了孤子脉冲的产生[93]、啁啾孤子脉冲的压缩[94, 95]和孤子脉冲的传输研究[96-105]，国内对光纤拉曼放大的研究大多集中在线性传输系统中的泵浦源数目、功率、波长等的选择和优化，对利用光纤拉曼放大的光孤子传输系统的研究较少[139-141]。目前，拉曼放大光孤子传输系统的研究中大多采用自相关技术等测量脉冲时域变化，难于准确判断脉冲的时域波形等[80, 93-98]。鉴于以往实验条件等各种因素的限制，采用能准确测量脉冲时域波形的 SHG-FROG 技术研究拉曼放大对光孤子传输特性的影响，并作相应的理论分析，将对拉曼放大器及其通信系统的进一步研究具有一定的指导意义。

近年来，模分复用研究由美国、日本和部分欧洲国家主导[251-261]。在中国，2014 年科技部资助了天津大学牵头的“多维复用光纤通信基础研究”(“973”基础研究项目[251])。天津大学、清华大学、北京大学、北京邮电大学、中国科学院半导体研究所、武汉邮电科学研究院、华中科技大学、北京交通大学先后加入模分复用领域的研究工作。总之，模分复用领域处于基础研究阶段，已经成为新一代通信系统的研究热点[251-261]。模分复用研究涉及模分复用光纤、模式产生与转换、复用解复用机制、模分复用光放大器、模分复用光传输系统、模分复用光网络等。本书主要在少模光纤、少模复用器、少模光纤系统中的模拟信号传输和少模光纤系统网络中的数字脉冲传输等方面[253, 254, 256-257, 260, 261]开展系列创新性工作。

1.3 光纤通信系统与脉冲传输理论

1.3.1 光纤通信系统中的关键光电子器件基础

光纤通信系统的大多数光电子器件(如光孤子光源、波分复用光纤通信系统的超连续谱光源、拉曼放大器、光逻辑器件等)使用了光纤，几乎都涉及脉冲在光纤中的传输等问题，研究脉冲传输特性对光纤通信系统及其光电子器件的设计和优化具有重要意义。本节对部分关键光电子器件的基础介绍有助于后续章节相关问题的深入研究。

1.3.1.1 光孤子光源

光孤子概念是由 Hasegawa[79]和 Tappert 于 1973 年首次提出的。当光纤的线性色散效应和非线性自相位调制效应达到平衡时，脉冲在光纤中可演化形成光孤子，时域呈双曲正割波形，其时域波形和脉宽在随后的传播过程中保持不变，可应用于光通信中。1980 年，Gouveia-Neto, Iwatsuki, Murphy 和 Mollenauer 等首次在实验中观察到了光孤子[86]。研究表明[80-84, 87-149]，光孤子通信是高速、长距离、大容量通信的优选方案，是 21 世纪最有潜力和应用前景的通信方式之一。光孤子光源是实现光孤子通信的关键光电子器件。目前，利用各种半导体激光器、光纤激光器等都能成功研制光孤子光源。光孤子光源在光纤通信、光谱学和生命科学等领域具有重要的应用价值，目前已成为光电子技术领域十分活跃的研究课题。本书以半导体光孤子光源为例介绍光孤子光源的基本原理。

半导体光孤子光源可以由增益开关半导体激光器、Fabry-Perot(F-P)腔滤波器和掺铒光纤放大器三大部分构成。增益开关半导体激光器产生可变速率光脉冲。改变激光器的偏置电流，可得到不同脉宽、谱宽和强度的脉冲。在增益开关状态下，会伴随较大的频率啁啾。适当设计 F-P 腔滤波器参数，使其谐振频率与信号中心频率一致，腔体带宽与信号带宽相等，可利用 F-P 腔滤波器有效滤除脉冲的啁啾成分。通常情况下，滤波后脉冲接近变换极限，其平均功率很低，远低于孤子阈值功率。随后，可以采用掺铒光纤放大器模块，对脉冲进行放大，以达到或超过孤子的阈值功率，从而使脉冲在光纤中传输演化成孤子。本书部分章节研究了啁啾脉冲的传输演化等问题。

1.3.1.2 波分复用光纤通信系统的超连续谱光源

随着信息化社会的发展，传输信息量的激增，实时性要求的提高，用户数量的扩大，传统的时分复用扩容方法已经不能满足需求，波分复用技术成为增加系统容量的有效方法。波分复用系统的常规光源仍是具有不同波长且波长间隔须精确控制的多个分离的激光器。随着波分复用信道数量的增加，系统常规光源出现了波长间隔控制困难、成本上升以及可靠性不高等问题。研究表明[80, 51, 52, 212-215]波分复用光纤通信系统的超连续谱光源具有宽带宽、低噪声、信道间的光频稳定和容易与光纤耦合等特点，是未来超大容量光纤通信系统最有希望的光源。

波分复用超连续谱光源主要由锁模光纤激光器、调制器、放大器、超连续谱光纤、分波器等部分构成。锁模光纤激光器能产生高速、稳定的短脉冲，作为种子脉冲源。调制器一般采用 $LiNbO_3$ 调制器，对种子脉冲进行调制。放大器通常采用掺铒光纤放大器，将调制后的种子脉冲放大到数百毫瓦至瓦级输出功率。超连续谱光纤常采用色散位移光纤和色散平坦光纤等，其作用是利用强非线性(如自相位调制)将入射的高功率脉冲展宽形成超连续光谱。分波器可以采用阵列波导型波分复用器等，将超连续光谱分割成许多按一定波长间隔排列的脉冲光源，供系统使用。本书部分章节研究了啁啾等参量对脉冲形成超连续谱的影响。

1.3.1.3 偏振复用与光逻辑器件

偏振复用是在同一传输信道同一波长脉冲中，利用两个相互正交的光偏振态独立地传输两路数据信号，偏振复用能够有效地降低脉冲间的相互作用，增大系统传输距离，并且能使每个信道的数据速率提高一倍，已经成为增大波分复用系统容量的新选择。

光逻辑器件的基本工作原理源自两个正交偏振脉冲俘获(束缚态)的非线性现象。在光逻辑器件中，一个偏振态脉冲作为数据信号，另一个偏振态脉冲作为控制信号，其初始时间间隔可以不同；在输入端，通过调节控制信号孤子和数据信号脉冲的时间间隔、啁啾等来控制两脉冲是否处于束缚态，使其产生不同的时间延迟；在输出端，可以通过设置相应的延迟来接收或不接收数据信号。

在偏振复用系统和光逻辑器件中都存在两个偏振脉冲的碰撞作用，其研究具有重要的实用价值[80]，本书在部分章节中研究了脉冲的碰撞特性。

1.3.1.4　光纤拉曼放大器

光纤拉曼放大器是基于受激拉曼散射原理，以光纤作为增益介质而实现的全光放大。相对于稀土掺杂的光纤放大而言,它具有更大的增益带宽(1270～1670nm)、灵活的增益谱区、更低的放大器自发辐射噪声以及能够有效抑制信噪比(SNR)劣化等优点，近年来在光纤传输系统中获得越来越多的应用[80, 93-105, 139-141]。

受激拉曼散射是非线性光纤光学中非常重要的非线性过程。光纤拉曼散射是入射光子与光纤分子相互作用时，入射光子发射或吸收一个声子(使光纤分子完成两个振动态之间的跃迁)，产生频率下移的斯托克斯(Stokes)光子或频率上移的反斯托克斯光子。光纤的拉曼声子频率为$\Delta\nu=1.32\times10^{13}$Hz，斯托克斯拉曼光子$\nu_s=\nu_p-\Delta\nu$，反斯托克斯拉曼光子$\nu_r=\nu_p+\Delta\nu$。其中$\nu_p$、$\nu_s$和$\nu_r$分别为泵浦光、斯托克斯光和反斯托克斯光的频率。拉曼放大主要是利用泵浦光产生频率下移的斯托克斯光波形成较大的拉曼增益来放大信号光波。

光纤拉曼放大器示意图如图 1-1 所示。光信号(optical signal)可以是连续光信号或者脉冲信号，WDM 是波分复用器，将光信号和泵浦激光器(pumping laser)产生的泵浦波耦合到光纤(fiber)中，利用拉曼增益效应放大光信号。实验中所用光纤可以根据实际应用作不同选择。泵浦激光器可以采用半导体激光器，分为前向泵浦、后向泵浦和双向泵浦。本书在部分章节中研究了拉曼放大对脉冲信号的影响。

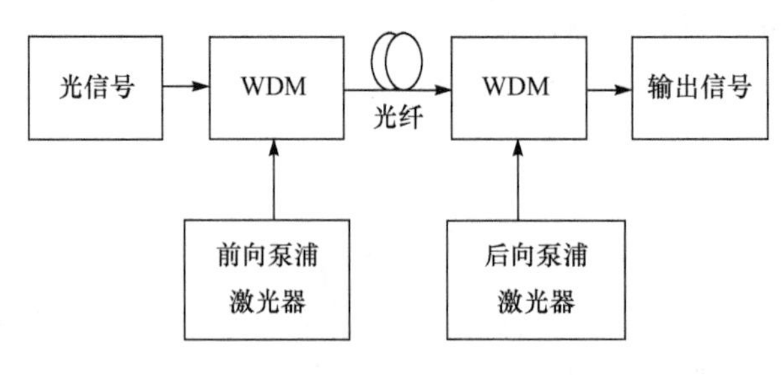

图 1-1　光纤拉曼放大器示意图

1.3.2　光纤的基本特性

1.3.2.1　光纤损耗

光纤损耗是光纤的重要特性之一，它将使传输的光信号产生衰减。当输入光功率和接收机灵敏度给定后，光纤损耗将成为决定系统无中继传输距离的重要因素。通常情况下，光信号在光纤中传输时，输出光功率随传输距离呈指数规律衰减[80]，即

$$P_L = P_0 \mathrm{e}^{-\alpha L} \tag{1-1}$$

式中，P_L是输出光功率，P_0是输入光功率，L是传输光纤长度，α是衰减系数，通常称为光纤损耗。光纤损耗一般通过下式用 dB/km 来表示，即

$$\alpha_{\mathrm{dB}} = -\frac{10}{L}\lg\left(\frac{P_L}{P_0}\right) = 4.343\alpha\ (\mathrm{dB/km}) \tag{1-2}$$

自 1979 年以来，在波长 1.55 μm 处的最小光纤损耗约为 0.2dB/km[33]。导致光纤损耗的因素非常复杂，降低光纤损耗主要依赖于光纤制造工艺的提高及对光纤材料的研究[80]。

1.3.2.2　光纤色散

1) 光纤的色散参量

在光纤中，色散是一种引起传输信号畸变的物理现象，它是由传输信号的不同频率成分和不同模式成分具有不同的群速度造成的。

当一束电磁波与电介质的束缚电子相互作用时，介质的响应通常与光波频率有关。色散的起源与介质通过束缚电子吸收电磁辐射的特征共振频率有关，远离介质共振频率时，折射率满足 Sellmeier 方程[80]：

$$n^2(\omega) = 1 + \sum_{j=1}^{m} \frac{B_j \omega_j^2}{\omega_j^2 - \omega^2} \tag{1-3}$$

式中，ω_j是谐振频率，B_j是j阶谐振强度。对于石英材料，相应参数为：$m=3$，B_1=0.6961663，B_2=0.4079426，B_3=0.8974794，λ_1=0.0684043μm，λ_2=0.1162414μm，λ_3=9.896161μm，$\lambda_j = 2\pi c/\omega_j$，$c$是真空中的光速。

由于不同的频率分量对应于不同的脉冲传输速度$c/n(\omega)$，因而色散在短脉冲传输中起关键作用。当非线性效应不明显时，由色散引起的脉冲变宽对光通信系统非常有害。在数学上，光纤的色散效应可以用在中心频率ω_0处展开成模传输常数β的泰勒级数来表示[80]：

$$\beta(\omega)=n(\omega)\frac{\omega}{c}=\beta_0+\beta_1(\omega-\omega_0)+\frac{1}{2}\beta_2(\omega-\omega_0)^2+\Lambda \tag{1-4}$$

式中

$$\beta_m=\left(\frac{\mathrm{d}^m\beta}{\mathrm{d}\omega^m}\right)_{\omega=\omega_0},\quad m=0,1,2,\cdots \tag{1-5}$$

参量 β_1、β_2 和折射率 n 有关，它们满足下面关系[80]：

$$\beta_1=\frac{n_{\mathrm{g}}}{c}=\frac{1}{v_{\mathrm{g}}}=\frac{1}{c}\left(n+\omega\frac{\mathrm{d}n}{\mathrm{d}\omega}\right) \tag{1-6}$$

$$\beta_2=\frac{1}{c}\left(2\frac{\mathrm{d}n}{\mathrm{d}\omega}+\omega\frac{\mathrm{d}^2n}{\mathrm{d}\omega^2}\right) \tag{1-7}$$

式中，n_{g} 是群折射率，v_{g} 是群速度，脉冲包络以群速度运动。参量 β_2 是群速色散(GVD)参量，是研究脉冲传输的重要参量之一，实际应用中无法直接测量 β_2 的数值，但是可以通过直接测量色散参量 D 并通过下式计算得到[80]：

$$D=\frac{\mathrm{d}\beta_1}{\mathrm{d}\lambda}=-\frac{2\pi c}{\lambda^2}\beta_2\approx-\frac{\lambda}{c}\frac{\mathrm{d}^2n}{\mathrm{d}\lambda^2} \tag{1-8}$$

根据色散参量 β_2 或 D 的符号，光纤中的非线性效应表现出显著不同的特征。$\beta_2=0$ 处的波长称为零色散波长 λ_D。若波长 $\lambda<\lambda_D$，光纤表现出正常色散，$\beta_2>0$。在正常色散区，脉冲的较高的频率分量(蓝移)比较低的频率分量(红移)传输得慢。$\beta_2<0$ 的所谓的反常色散区情况正好相反。由于在反常色散区通过色散和非线性效应之间的平衡，光纤能维持光孤子，人们在非线性效应的研究中，对反常色散区特别感兴趣。作者主要研究了啁啾脉冲在反常色散区的传输特性。

色散的一个重要特性是，由于群速度失配，不同波长下的脉冲在光纤内以不同的速度传输，这一特性导致了走离效应，它在涉及两个或更多个交叠脉冲的非线性现象的描述中起了重要的作用。更准确地说，当传输得较快的脉冲完全通过传输得较慢的脉冲后，两脉冲之间的互作用将停止。作者将在后续章节的脉冲在双折射光纤中的碰撞特性中讨论。

2) 光纤的分类

常规单模光纤(如 G.652 光纤)，零色散波长 $\lambda_D=1.31\mu\mathrm{m}$。根据色散参量 D(或 β_2)依赖于光纤设计参数(如芯径等)的特性可以将零色散波长 λ_D 移到具有最小损耗的 1.55μm 附近，称为色散位移光纤。根据在 1.55μm 处色散参量 D 是否为零，色散位移光纤可以分别称为零色散位移光纤和非零色散位移光纤(如 G.655 光纤)，这些光纤已经商用化。在 1.6μm 以上区域，有些光纤表现出具有较大的正 β_2 值的群速度色散，这种光纤称为色散补偿光纤(DCF)。用于 WDM 系统的小色散斜率光纤在最近几年得到发展。在较大波长范围内(1.3～1.6μm)，可以利用多包层实现具有低色散值的色散平坦光纤。通过沿光纤逐渐减小纤芯直径，可以制造出色散渐减光纤。

1.3.2.3 光纤的非线性特性

在高强度电磁场中，光纤同任何电介质一样对光的响应都会变成非线性。从其基能级看，介质非线性响应的起因与施加到它上的场影响下的束缚电子的非谐振运动有关，结果导致电偶极子的极化强度 $\vec{P}$ 对于电场 $\vec{E}$ 是非线性的，但满足通常的关系式[80]

$$\vec{P} = \varepsilon_0(\chi^{(1)} \cdot \vec{E} + \chi^{(2)} : \vec{E}\vec{E} + \chi^{(3)} \vdots \vec{E}\vec{E}\vec{E} + \cdots) \tag{1-9}$$

式中，ε_0 是真空中的介电常数，$\chi^{(j)}$ (j=1，2，…)为 j 阶电极化率，考虑到光的偏振效应，$\chi^{(j)}$ 是 j+1 阶张量。线性电极化率 $\chi^{(1)}$ 对 $\vec{P}$ 的贡献是主要的，它的影响包含在折射率 n 和衰减常数 α 中。二阶电极化率 $\chi^{(2)}$ 对应于二次谐波的产生、和频运转等非线性效应。然而，$\chi^{(2)}$ 只在某些分子结构非反演对称的介质中才不为零。SiO_2 分子是对称结构，因而对于石英光纤 $\chi^{(2)}=0$，光纤通常不显示二阶非线性效应，然而电四极矩和磁偶极矩能产生弱的二阶非线性效应，纤芯中的缺陷和色心在特定条件下也对二次谐波的产生有影响。

1) 非线性折射率

光纤中的最低阶非线性效应起源于三阶电极化率 $\chi^{(3)}$，它是引起诸如三次谐波产生、四波混频以及非线性折射等现象的主要原因。然而，除非采取特别的措施实现相位匹配，牵涉到新频率产生的(三次谐波的产生或四波混频)非线性过程在光纤中是不易发生的。因而，光纤中的大部分非线性效应起源于非线性折射率，而折射率与光强有关的现象是由 $\chi^{(3)}$ 引起的。

2) 受激非弹性散射

三阶电极化率 $\chi^{(3)}$ 导致的非线性效应是弹性的，因为在电磁场和电介质之间无能量交换。光纤中有两个重要的非线性效应起因于光场把部分能量传递给介质，属于受激非弹性散射，它们都和石英的振动激发态有关，这就是众所周知的受激拉曼散射(SRS)和受激布里渊散射(SBS)。二者的主要差别是：在 SRS 中参与的是光学声子，而在 SBS 中参与的是声学声子。一个入射场的光子的湮灭，产生了一个下移斯托克斯频率的光子和保持能量与动量守恒的另一个具有相应能量与动量的声子。当然，如果能吸收一个具有相应能量和动量的声子，也可能产生有更高能量的光子，称为反斯托克斯频率。尽管 SRS 和 SBS 在起因上非常相似，但由于声子和光子不同的色散关系，在光纤中的 SBS 只发生在后向，而 SRS 在两种方向均能发生。

1.3.2.4 光纤中的传输模式

对任何频率 ω，光纤仅能容纳有限的几个传输模式，这些传输模式的电场强度 $\tilde{E}(\vec{r},\omega)$ 的空间分布是满足一定边界条件的波动方程的解。已经证明存在两类光纤模式，称为 HE_{mn} 和 EH_{mn}。对 $m=0$，由于其电场和磁场的轴向分量趋于零，这些模式类似于平面波导的横电场模(TE)和横磁场模(TH)；然而，当 $m>0$ 时，光纤模是混合型的，即电磁场的所有六个分量均不为零。

在给定波长的情况下，特定光纤所容纳的模式数依赖于其设计参数，即纤芯半径 a 和纤芯-包层折射率差 n_1-n_2。单模光纤是在给定工作波长上，只传输基模 $LP_{01}(HE_{11})$ 的光波导。近年来，大多数通信系统都使用单模光纤，本书主要研究单模传输问题。

光纤的归一化频率满足[80]

$$V=\frac{2\pi}{\lambda}a\sqrt{n_1^2-n_2^2} \tag{1-10}$$

实现单模传输的条件为：当

$$V<V_c=2.045 \tag{1-11}$$

时，光纤只能传输一个基模 $LP_{01}(HE_{11})$，其中 $V_c=2.045$ 是单模光纤的归一化截止频率。由式(1-10)和式(1-11)得到

$$\lambda_c=\frac{2\pi}{V_c}a\sqrt{n_1^2-n_2^2} \tag{1-12}$$

式中，λ_c 称为单模光纤的截止波长。当光纤剖面结构参数确定后，只有工作波长大于 λ_c 这一特定波长时，才能实现单模传输。

1.3.3 脉冲的基本传输方程

同所有的电磁现象一样，光纤中脉冲的传输服从麦克斯韦方程组。在国际单位制中，该方程组可写成[80]

$$\nabla\times\vec{E}=-\frac{\partial\vec{B}}{\partial t} \tag{1-13a}$$

$$\nabla\times\vec{H}=\vec{J}+\frac{\partial\vec{D}}{\partial t} \tag{1-13b}$$

$$\nabla\cdot\vec{D}=\rho_f \tag{1-13c}$$

$$\nabla\cdot\vec{B}=0 \tag{1-13d}$$

式中，$\vec{E}$、$\vec{H}$ 分别为电场强度矢量和磁场强度矢量；$\vec{D}$、$\vec{B}$ 分别为电位移矢量和磁感应强度矢量；电流密度矢量 $\vec{J}$ 和电荷密度 ρ_{f} 表示磁场的源。在光纤这样无自由电荷的介质中，显然 $\vec{J}=0$，$\rho_{\mathrm{f}}=0$。介质内传输的电磁场强度 $\vec{E}$ 和 $\vec{H}$ 增大时，电位移矢量 $\vec{D}$ 和磁感应强度 $\vec{B}$ 也随之增大，它们的关系通过物质方程联系起来：

$$\vec{D}=\varepsilon_0\vec{E}+\vec{P} \tag{1-13e}$$

$$\vec{B}=\mu_0\vec{H}+\vec{M} \tag{1-13f}$$

式中，ε_0 为真空中介电常数，μ_0 为真空中的磁导率，$\vec{P}$、$\vec{M}$ 分别为感应电极化强度和磁极化强度，在光纤这样的无磁性介质中 $\vec{M}=0$。

为完整表达光纤中光波的传输，需要找到电极化强度 $\vec{P}$ 和电场强度 $\vec{E}$ 的关系，当光频与介质共振频率接近时，$\vec{P}$ 的计算必须采用量子力学方法。但是在远离介质共振频率处，$\vec{P}$ 和 $\vec{E}$ 的关系满足公式(1-9)，引人关注的 0.5～2μm 波段内光纤的非线性正是这种情况。如果只考虑与 $x^{(3)}$ 有关的非线性效应，则感应电极化强度由线性和非线性两部分组成[80]

$$\vec{P}(\vec{r},t)=\vec{P}_{\mathrm{L}}(\vec{r},t)+\vec{P}_{\mathrm{NL}}(\vec{r},t) \tag{1-14}$$

式中，两部分的普适应关系分别为

$$\vec{P}_{\mathrm{L}}(\vec{r},t)=\varepsilon_0\int_{-\infty}^{\infty}x^{(1)}(t-t')\cdot E(\vec{r},t')\mathrm{d}t' \tag{1-15a}$$

$$\vec{P}_{\mathrm{NL}}(\vec{r},t)=\varepsilon_0\iiint x^{(3)}(t-t_1,t-t_2,t-t_3)\vdots\vec{E}(\vec{r},t_1)\vec{E}(\vec{r},t_2)\vec{E}(\vec{r},t_3)\mathrm{d}t_1\mathrm{d}t_2\mathrm{d}t_3 \tag{1-15b}$$

为求解麦克斯韦方程组，须作几个假设进行简化。首先，把 $\vec{P}_{\mathrm{NL}}$ 处理成 $\vec{P}_{\mathrm{L}}$ 的微扰，实际上，折射率的非线性变化小于 10^{-6}；其次，假定光场沿光纤长度方向其偏振态不变，因而其标量近似有效，事实并非如此，除非采用保偏光纤，但这种近似非常有效；最后，假定光场是准单色的，即对中心频率为 ω_0 的频谱，其谱宽为 $\Delta\omega$，且 $\Delta\omega/\omega_0\ll 1$。因为 ω_0 约为 10^{15}Hz，最后一项假定对脉宽大于等于 0.1ps 的脉冲是成立的。在慢变包络近似下，利用变量分离法求解麦克斯韦方程组得到：

(1) 脉冲在单模光纤中基模 HE_{11} 的模分布与高斯分布近似吻合[80]

$$F(x,y)=\exp[-(x^2+y^2)/w^2] \tag{1-16}$$

在特定的 V=2.4 情况下，实际模场分布与拟合高斯分布的比较，符合得很好；特别是在 V=2 附近有模场半径 $w=a$，表明对 V=2 的通信光纤，纤芯半径 a 和 w 基本一致。

(2) 脉冲在单模光纤中的基本传输方程为[80]

$$\frac{\partial A}{\partial z}=\mathrm{i}\sum_{n=1}^{\infty}\frac{\mathrm{i}^n\beta_n}{n!}\frac{\partial^n A}{\partial t^n}+\mathrm{i}\gamma|A|^2A-\frac{\alpha}{2}A \tag{1-17}$$

式中，A 为脉冲慢变包络电场 $A(z,t)$，z 为脉冲在光纤中传输的方向，t 为以中心波长群速度为移动参考系的时间参量。右边第一项为色散项，β_n 为各阶群速度色散系数，n 表示色散阶数；右边第二项表示脉冲传输中的非线性项，具体为自相位调制效应；最后一项表示光纤损耗项，α 为光纤损耗系数。非线性系数定义为

$$\gamma=\frac{n_2\omega_0}{cA_{\text{eff}}} \tag{1-18}$$

n_2 为非线性折射率系数，ω_0 为脉冲中心角频率，c 为真空中的光速，A_{eff} 为纤芯有效横截面积。A_{eff} 定义为

$$A_{\text{eff}}=\frac{\left(\iint|F(x,y)|^2\mathrm{d}x\mathrm{d}y\right)^2}{\iint|F(x,y)|^4\mathrm{d}x\mathrm{d}y}=\pi a^2 \tag{1-19}$$

尽管传输方程(1-17)已成功地解释了许多非线性效应,但它仍然需要根据实验等情况来改进。例如，方程(1-17)没有包含 SRS 和 SBS 的受激非弹性散射。若入射脉冲的峰值功率超过其阈值，就必须考虑 SRS 和 SBS 效应。它们会把泵浦能量传递给与泵浦脉冲同向或反向共同传输的斯托克斯脉冲，通过拉曼增益或是布里渊增益及交叉相位调制(XPM)产生相互作用。方程(1-17)对脉宽小于 5ps 且包含多个光学周期的超短脉冲也需要改进。已证明最重要的限制是忽略了脉冲内的拉曼散射。脉冲在光纤内的传输过程中,脉冲内拉曼散射使脉冲频谱移向红光一侧，这种现象称为自频移。考虑了脉冲内拉曼散射效应和脉冲自变陡效应，引入以群速度 v_{g} 移动的延时参考系

$$T=t-z/v_{\mathrm{g}}\equiv t-\beta_1 z \tag{1-20}$$

且对高阶非线性项作了泰勒级数近似后，得到[80]

$$\frac{\partial A}{\partial z}=\mathrm{i}\sum_{n=2}^{\infty}\frac{\mathrm{i}^n\beta_n}{n!}\frac{\partial^n A}{\partial T^n}+\mathrm{i}\gamma\left[|A|^2A+\frac{\mathrm{i}}{\omega_0}\frac{\partial(|A|^2A)}{\partial T}-T_{\mathrm{R}}A\frac{\partial|A|^2}{\partial T}\right]-\frac{\alpha}{2}A \tag{1-21}$$

方程(1-21)比(1-17)的非线性项多出了两部分，非线性项的第二部分为脉冲自变陡效应，第三部分为脉冲内拉曼散射效应，T_{R} 为脉冲内拉曼散射系数，与拉曼增益的斜率有关，通常为 3fs。方程(1-21)即为脉冲在光纤中传输所满足的广义非线性薛定谔方程，非线性薛定谔方程是非线性科学的一个基本方程，并被广泛用于研究皮秒或飞秒脉冲的传输等,本书部分工作就是在方程(1-21)基础上展开研究的。另外，该方程能够以不同形式应用于光学领域。

单模光纤实际上并非真正意义上的单模，因为它能维持具有相同空间分布的两正交偏振模式。方程(1-21)忽略了偏振的影响，或者说它们适用于入射脉冲沿光纤一主轴线性偏振的情况。

在理想光纤中(光纤在整个长度上保持严格的圆柱对称性)，这两种模式是简并的，或者说它们的有效折射率相等。实际上，由于沿光纤存在纤芯形状的意外改变和各向异性应力，所有光纤均表现出一定程度的模式双折射(即 $n_x \neq n_y$)，而且模式双折射度 $B_m = |n_x - n_y|$ 沿 x 轴和 y 轴的取向在约 10m 的长度上就会随机改变，除非采取特殊的预防措施。在保偏光纤中，施加的固有双折射要比由应力和纤芯形状变化引起的随机双折射大得多，结果保偏光纤在整个长度上其双折射几乎是常数，这种双折射称为线性双折射。当光纤中的非线性效应变得重要时，足够强的光场能引起非线性双折射，其大小与光场强度有关。这种偏振效应最早于 1964 年在体或块非线性介质中观察到，随后，人们对此进行了广泛的研究。

具有恒定模式的双折射光纤有两个主轴，若脉冲沿这两个主轴入射，光纤能保持其线性偏振态。根据偏振光沿这两个主轴的传播速度不同，可分别称之为慢轴和快轴，假定 $n_x > n_y$，则 n_x 和 n_y 分别是沿慢轴和快轴的折射率。在慢变包络近似下，可以得到具有正交偏振分量的两个孤子在双折射光纤中的归一化耦合模方程[80]

$$\mathrm{i}\left(\frac{\partial u}{\partial \xi} + \delta \frac{\partial u}{\partial \tau}\right) + \sum_{n=2}^{\infty} \frac{\mathrm{i}^n \beta_n}{n!|\beta_2| T_0^{n-2}} \frac{\partial^n u}{\partial \tau^n} + (|u|^2 + B|v|^2)u + \frac{\mathrm{i}}{2}\Gamma u = 0 \tag{1-22a}$$

$$\mathrm{i}\left(\frac{\partial v}{\partial \xi} - \delta \frac{\partial v}{\partial \tau}\right) + \sum_{n=2}^{\infty} \frac{\mathrm{i}^n \beta_n}{n!|\beta_2| T_0^{n-2}} \frac{\partial^n v}{\partial \tau^n} + (|v|^2 + B|u|^2)v + \frac{\mathrm{i}}{2}\Gamma u = 0 \tag{1-22b}$$

式中，u 和 v 分别是沿慢、快轴的脉冲归一化慢变包络电场，$\Gamma = \alpha L_D = \alpha T_0^2 / |\beta_2|$ 与 $\delta = (\beta_{1x} - \beta_{1y})T_0 / 2|\beta_2|$ 反映了两个孤子的群速度失配，归一化时间 $\tau = (t - \overline{\beta}_{1z}) / T_0$，其中 $\overline{\beta}_1 = 0.5(\beta_{1x} + \beta_{1y})$ 与脉冲的平均群速度成反比。本书利用方程(1-22)数值研究了啁啾正交偏振孤子间的碰撞特性，为光逻辑器件和偏振复用系统的应用提供了重要理论依据。

1.3.4 脉冲传输的数值计算方法

方程(1-21)是非线性偏微分方程，在一般情况下不适于解析求解，除非是在能使用逆散射方法的某些特殊情况下才有可能。为了研究光纤中脉冲的传输特性，一般采用数值计算的方法。本节以求解方程(1-21)为例，介绍了分步傅里叶数值计算方法的基本原理，这种方法相对于大多数有限差分法有较快的速度，部分原因是采用了快速傅里叶变换(FFT)算法。

为了简单说明分步傅里叶方法的基本原理，把方程(1-21)改成如下形式[80]：

$$\frac{\partial A}{\partial z}=(\hat{L}+\hat{N})A \tag{1-23}$$

式中，

$$\hat{L}=\mathrm{i}\sum_{n=2}^{\infty}\frac{\mathrm{i}^n\beta_n}{n!}\frac{\partial^n}{\partial T^n}-\frac{\alpha}{2} \tag{1-24}$$

$\hat{L}$ 是差分算符，它表示线性介质的色散和吸收。

$$\hat{N}=\mathrm{i}\gamma\left[|A|^2+\frac{\mathrm{i}}{\omega_0}\frac{1}{A}\frac{\partial(|A|^2A)}{\partial T}-T_{\mathrm{R}}\frac{\partial|A|^2}{\partial T}\right] \tag{1-25}$$

$\hat{N}$ 是非线性算符，它决定了脉冲传输过程中光纤的非线性效应。

一般来说，沿光纤的长度方向，色散和非线性是同时作用的。分步傅里叶方法通过假定在传输过程中，光场每通过一小段距离 h，色散和非线性效应可分别作用，得到近似结果。更准确地说，从 z 到 $z+h$ 的传输过程中分两步进行。第一步，仅有色散作用，方程(1-23)中的非线性作用为零(即 $\hat{N}=0$)；第二步，仅有非线性作用，方程(1-23)中的色散作用为零(即 $\hat{L}=0$)。方程(1-23)可以表示为[80]

$$A(z+h,T)=\exp[(\hat{L}+\hat{N})\cdot h]\,A(z,T) \tag{1-26}$$

根据两个非对易算符的 Bake-Hausdorff 公式，忽略算符 $\hat{L}$ 和 $\hat{N}$ 的非对易性，方程(1-26)近似为

$$A(z+h,T)\approx\exp(h\hat{L})\exp(h\hat{N})A(z,T) \tag{1-27}$$

或者采用

$$A(z+h,T)\approx\exp(0.5\cdot h\hat{L})\exp(h\hat{N})\ \exp(0.5\cdot h\hat{L})A(z,T) \tag{1-28}$$

其中

$$\exp(h\hat{L})A(z,T)=F_{\mathrm{T}}^{-1}\{\exp[h\hat{L}(\mathrm{i}\omega)]\,F_{\mathrm{T}}A(z,T)\} \tag{1-29}$$

式中，F_{T} 表示傅里叶变换，$h\hat{L}(\mathrm{i}\omega)$ 从方程(1-24)通过 $\mathrm{i}\omega$ 代替微分算符 $\partial/\partial T$ 得到，ω 为傅里叶域中的频率，F_{T}^{-1} 表示傅里叶逆变换。

分步傅里叶方法已广泛应用于各种光学领域，包括大气中的光传输、折射率梯度光纤、半导体激光器、非稳腔及波导耦合器等。虽然用此方法运算速度快，相对简洁，但需要小心选择 z 和 T 的步长，以保证精度要求。

1.4 本章小结

本章主要论述了啁啾脉冲传输中的若干关键技术研究现状、存在的主要问题，介绍了光纤通信系统中的关键光电子器件基础、传输光纤特性，讨论了脉冲传输理论基础与数值计算方法。

参考文献

[1] Maiman T H. Stimulated optical radiation in ruby. Nature, 1960, 187 (4736): 493～494.

[2] Zhang F, Yao J Q, Hou N, et al. The method research of used laser to processes the slot pipe applied to petroleum exploitation. Proc. of SPIE, 2002, 4915: 296～297.

[3] Hallada M R, Walter R F, Seiffert S L. High power laser rock cutting and drilling in mining operations: Initial feasibility tests. Proc. of SPIE, 2001, 4184: 590～593.

[4] Lambert R W, Cortés-Martínez R, Waddie A J, et al. Compact optical system for pulse-to-pulse laser beam quality measurement and applications in laser machining. Applied Optics, 2004, 43(26): 5037～5046.

[5] Imai K, Takahashi S, Kamahori M, et al. Multi-capillary DNA equencer. Hitachi Review, 1999, 48(3): 107～109.

[6] Abdula R M, Saleh B E A. Dynamic spectra of pulsed laser diodes and propagation in single-mode fibers. IEEE J Quantum Electronics, 1986, QE-22 (11): 2123～2130.

[7] Iwatsuki K, Takada A, Saruwatari M. Optical soliton propagation using 3GHz gain-switched 1.3 μm laser diodes. Electronics Letters, 1988, 24(25): 1572～1574.

[8] Dudley J M, Harvey J D, Leonhardt R. Coherent pulse propagation in a mode-locked argon laser. J. Opt. Soc. Am. B, 1993, 10(5): 840～851.

[9] Akhmediev N, Soto-Crespo J M. Propagation dynamics of ultrashort pulses in nonlinear fiber couplers. Physical Review E, 1994, 49(5): 4519～4529.

[10] Peeters M, Verschaffelt G, Speybrouck J, et al. Propagation of spatially partially coherent emission from a vertical-cavity surface-emitting laser. Optics Letters, 2006, 31(9): 1178～1180.

[11] Yu Y, Lu Y, Liu L, et al. Experimental Demonstration of Single Carrier 400G/500G in 50-GHz Grid for 1000-km Transmission. Optical Fiber Communication Conference, OSA Technical Digest (online) (Optical Society of America, 2017), paper Tu2E.4.

[12] Gao G, Luo M, Li X, et al. Transmission of 2.86 Tb/s data stream in silicon subwavelength grating waveguides. Opt. Express, 2017, 25: 2918～2927.

[13] Zou K H, Zhu Y X, Zhang F. 800Gb/s (8 × 100Gb/s) Nyquist half-cycle single-sideband modulation direct-detection transmission over 320 km SSMF at C-band. J. Lightwave Technol, 2017, 35: 1900～1905.

[14] Fermann M E, Yang L M, Stock M L, et al. Environmentally stable Kerr-type mode-locked erbium fiber laser producing 360 fs pulses. Optics Letters, 1994, 19(l) :43～45.

[15] Galvanauskas A, Fermann M E, Harter D, et al. All-fiber femtosecond pulse amplification circuit using chirped Bragg gratings. Applied Physics Letters, 1995, 66(9): 1053～1055.

[16] Yang L M, Sosnowski T, Stock M L, et al. Chirped-pulse amplification of ultrashort pulses with a multimode Tm^{3+} : ZBLAN fiber upconversion amplifier. Optics Letters, 1995, 20(9):1044～1046.

[17] Walton D T, Nees J, Mourou G. Broad-bandwidth pulses amplification to the 10μJ level in an Yb^{3+}-doped germanosilicate fiber. Optics Letters, 1996, 21(14): 1061～1063.

[18] Lefort L, Price J H V, Richardson D J, et al. Practical low-noise stretched-pulse Yb^{3+} doped fiber laser. Optics Letters, 2002, 27(5): 291～293.

[19] Bartels R A, Paul A, Green H, et al. Generation of spatially coherent light at extreme ultraviolet wavelengths. Science, 2002, 297 (5580): 376～378.

[20] Shiraishi T, Mori M, Kondo K. Estimation of the pulse width of X-ray emission from Xe clusters excited by a subpicosecond intense Ti: Sapphire laser pulse. Phys. Rev. A, 2002, 65 (4): 045201～045204.

[21] Gonoskov A, Tsatrafyllis N, Kominis I K, et al. Quantum optical signatures in strong-field laser physics: Infrared photon counting in high-order-harmonic generation. Scientific Reports, 2016, 6: 32821.

[22] Gonoskov A, Bastrakov S, Efimenko E, et al. Extended particle-in-cell schemes for physics in ultrastrong laser fields: Review and developments. Phys. Rev. E, 2015, 92: 023305.

[23] Krüger M, Thomas S, Breuer John, et al. Nanooptics and electrons: From strong-field physics at needle tips to dielectric laser acceleration. 2014, 27th International Vacuum Nanoelectronics Conference (IVNC).

[24] 盛政明, 等. 强场激光物理研究前沿. 上海: 上海交通大学出版社, 2014.

[25] Metz W D. A new approach to thermonuclear power. Science, 1972, 177(4055): 1180～1182.

[26] Lubin M J. Laser fusion research. Science, 1975, 187(1478): 701～702.

[27] Speck D, Bliss E, Glaze J, et al. The Shiva laser-fusion facility. IEEE J. Quant. Electron, 1981, QE-17(9): 1599～1619.

[28] Normile D. National laboratories: Laser fusion with a fast twist. Science, 1997, 275(5304): 1254.

[29] Service R F. Physics: Laser labs race for the petawatt. Science, 2003, 301 (5630): 154～156.

[30] Paoloni C, Gamzina D, Himes L, et al. THz backward-wave oscillators for plasma diagnostic in nuclear fusion. IEEE Transactions on Plasma Science, 2016, 44(4): 369～376.

[31] Betti R. Status and prospects for burning plasmas via laser fusion. 2016 IEEE International Conference on Plasma Science (ICOPS).

[32] Picciotto A, Margarone D, Velyhan A, et al. Boron-proton nuclear-fusion enhancement induced in boron-doped silicon targets by low-contrast pulsed laser. Hys. Rev. X, 2014, 4: 031030.

[33] Miya T, Terunuma Y, Hosaka T, et al. Ultimata low-loss single mode fiber at 1.55 mm. Electronics Letters, 1979, 15(4): 106～108.

[34] Desurvire E. Analysis of noise figure spectral distribution in erbium doped fiber amplifiers pumped near 980 and 1480 nm. Applied Optics, 1990, 29 (21): 3118～3125.

[35] Quimby R S. Output saturation in a 980-nm pumped erbium-doped fiber amplifier. Applied Optics, 1991, 30(18): 2546～2552.

[36] Georges T, Delevaque E. Analytic modeling of high-gain erbium-doped fiber amplifiers. Optics Letters, 1992, 17(16): 1113～1115.

[37] Wen S. Distributed erbium-doped fiber amplifier for soliton transmission. Optics Letters, 1994, 19(1): 22～24.

[38] Bertilsson K, Andrekson P A, Olsson B E. Noise figure of erbium doped fiber amplifiers in the saturated regime. IEEE Photonics Technology Letters, 1994, 6(2): 199～201.

[39] Achtenhagen M, McElhinney M, Nolan S. High-power 980-nm pump laser modules for erbium-doped fiber amplifiers. Applied Optics, 1999, 38(27): 5765～5767.

[40] Sohn I B, Song J W. Gain flattened and improved double-pass two-stage EDFA using microbending long-period fiber gratings. Optics Communications, 2004, 236: 141～144.

[41] Cowle G. The State of the Art of Modern Non-SDM Amplification Technology in Agile Optical Networks: EDFA and Raman Amplifiers and Circuit Packs//Optical Fiber Communication Conference, OSA Technical Digest (online) (Optical Society of America, 2017), paper M3G.1.

[42] Huang Y, Cho P B, Samadi P, et al Dynamic Power Pre-adjustments with Machine Learning that Mitigate EDFA Excursions during Defragmentation//Optical Fiber Communication Conference, OSA Technical Digest (online) (Optical Society of America, 2017), paper Th1J.2.

[43] Liao P, Bao C, Kordts A, et al. Experimental Investigation of the Effect of EDFA-Generated ASE Noise added to the Pump of a Kerr Frequency Comb//Optical Fiber Communication Conference, OSA Technical Digest (online) (Optical Society of America, 2017), paper M3F.

[44] Huang Y S, Gutterman C L, Samadi P, et al. Dynamic mitigation of EDFA power excursions with machine learning. Opt. Express, 2017, 25: 2245～2258.

[45] Liang X J, Kumar S. Optical back propagation for fiber optic networks with hybrid EDFA Raman amplification. Opt. Express, 2017, 25: 5031～5043.

[46] Lundberg L, Andrekson P A, Karlsson M. Power consumption analysis of hybrid EDFA/Raman amplifiers in long-haul transmission systems. J. Lightwave Technol, 2017, 35: 2132～2142.

[47] Eznaveh Z S, Fontaine N K, Chen H, et al. Ultra-Low DMG Multimode EDFA//Optical Fiber Communication Conference, OSA Technical Digest (online) (Optical Society of America, 2017), paper Th4A.4.

[48] Wada M, Sakamoto T, Aozasa S, et al. Coupled 2-LP 6-core EDFA with 125 μm cladding diameter//Optical Fiber Communication Conference, OSA Technical Digest (online) (Optical Society of America, 2017), paper Th4A.6.

[49] 原荣. 光纤通信. 北京: 电子工业出版社, 2002.

[50] 高建平. 光纤通信. 西安: 西北工业大学出版社, 2005.

[51] Sotobayashi H, Chujo W, Konishi A, et al. Wavelength-band generation and transmission of 3.24-Tbit/s (81-channel×40-Gbit/s) carrier -suppressed return-to-zero format by use of a single supercontinuum source for frequency standardization. J. Opt. Soc. Am. B, 2002, 19(11): 2803～2809.

[52] Ohara T, Takara H, Yamamoto T, et al. Over-1000-channel ultradense wdm transmission with supercontinuum multicarrier source. Journal of Lightwave Technology, 2006, 24(6): 2311～2317.

[53] Gao F, Zhou S W, Li X, et al. 2 × 64 Gb/s PAM-4 transmission over 70 km SSMF using O-band 18G-class directly modulated lasers (DMLs). Opt. Express, 2017, 25: 7230～7237.

[54] Santa M D, Antony C, Power M, et al. 25Gb/s PAM4 Burst-Mode System for Upstream Transmission in Passive Optical Networks//Optical Fiber Communication Conference, OSA Technical Digest (online) (Optical Society of America, 2017), paper M3H.7.

[55] Eiselt N, Griesser H, Eiselt M H, et al. Real-Time 200 Gb/s (4×56.25 Gb/s) PAM-4 Transmission over 80 km SSMF using Quantum-Dot Laser and Silicon Ring-Modulator//Optical Fiber Communication Conference, OSA Technical Digest (online) (Optical Society of America, 2017), paper W4D.3.

[56] Zhu Y X, Zou K H, Chen Z Y, et al. 224 Gb/s optical carrier-assisted Nyquist 16-QAM half-cycle single-sideband direct detection transmission over 160 km SSMF. J. Lightwave Technol, 2017, 35: 1557～1565.

[57] Yoshida M, Nitta J, Kimura K, et al. Single-Channel 3.84 Tbit/s, 64 QAM Coherent Nyquist Pulse Transmission over 150 km with Frequency-Stabilized and Mode-Locked Laser//Optical Fiber Communication Conference, OSA Technical Digest (online) (Optical Society of America, 2017), paper Th2A.52.

[58] Kan T, Kasai K, Yoshida M, et al. 42.3-Tbit/s, 18-Gbaud 64QAM WDM Coherent Transmission of 160 km over Full C-band using an Injection Locking Technique with a Spectral Efficiency of 9 bit/s/Hz//Optical Fiber Communication Conference, OSA Technical Digest (online) (Optical Society of America, 2017), paper Th3F.5.

[59] Zhang J, Yu J, Chien H. 1.6Tb/s (4×400G) Unrepeatered Transmission over 205-km SSMF Using 65-GBaud PDM-16QAM with Joint LUT Pre-Distortion and Post DBP Nonlinearity Compensation//Optical Fiber Communication Conference, OSA Technical Digest (online) (Optical Society of America, 2017), paper Th2A.51.

[60] Hoang T M, Sowailem M, Osman M, et al. 280-Gb/s 320-km Transmission of Polarization-Division Multiplexed QAM-PAM with Stokes Vector Receiver//Optical Fiber Communication Conference, OSA Technical Digest (online) (Optical Society of America, 2017), paper W3B.4.

[61] Soma D, Wakayama Y, Igarashi K, et al. Partial MIMO-based 10-Mode-Multiplexed Transmission over 81km Weakly-coupled Few-mode Fiber//Optical Fiber Communication Conference, OSA Technical Digest (online) (Optical Society of America, 2017), paper M2D.4.

[62] Jung Y M, Kang Q Y, Zhou H Y, et al. Low-loss 25.3 km few-mode ring-core fiber for mode-division multiplexed transmission. J. Lightwave Technol, 2017, 35: 1363～1368.

[63] Huang A, Zacarias J C A, Fontaine N K, et al. 10-Mode Photonic Lanterns Using Low-Index Micro-structured Drilling Preforms//Optical Fiber Communication Conference, OSA Technical Digest (online) (Optical Society of America, 2017), paper Tu3J.5.

[64] Yerolatsitis S, Harrington K, Thomson R, et al. Mode-selective Photonic Lanterns from

Multicore Fibres//Optical Fiber Communication Conference, OSA Technical Digest (online) (Optical Society of America, 2017), paper Tu3J.6. (6-10mode).

[65] Zacarias J C A, Huang B, Fontaine N K, et al. Experimental analysis of the modal evolution in photonic lanterns//Optical Fiber Communication Conference, OSA Technical Digest (online) (Optical Society of America, 2017), paper Tu2J.7.

[66] Liu H, Wen H, Zacarias J C A, et al. 3×10 Gb/s Mode Group-Multiplexed Transmission over a 20 km Few-Mode Fiber Using Photonic Lanterns//Optical Fiber Communication Conference, OSA Technical Digest (online) (Optical Society of America, 2017), paper M2D.5.

[67] Benyahya K, Simonneau C, Ghazisaeidi A, et al. 5Tb/s transmission over 2.2 km of multimode OM2 fiber with direct detection thanks to wavelength and mode group multiplexing//Optical Fiber Communication Conference, OSA Technical Digest (online) (Optical Society of America, 2017), paper M2D.2.

[68] Ingerslev K, Gregg P, Galili M, et al. 12 Mode, MIMO-Free OAM Transmission//Optical Fiber Communication Conference, OSA Technical Digest (online) (Optical Society of America, 2017), paper M2D.1.

[69] Feng F, Jin X, O'Brien D, et al. Mode-Group Multiplexed Transmission using OAM modes over 1km Ring-Core Fiber without MIMO Processing//Optical Fiber Communication Conference, OSA Technical Digest (online) (Optical Society of America, 2017), paper Th2A.43.

[70] Chen Z, Hefferman G, Wei T. Digitally controlled chirped pulse laser for sub-terahertz-range fiber structure interrogation. Opt. Lett., 2017, 42: 1007～1010.

[71] Pastor-Graells J, Cortés L R, Fernández-Ruiz M R, et al. SNR enhancement in high-resolution phase-sensitive OTDR systems using chirped pulse amplification concepts. Opt. Lett., 2017, 42: 1728～1731.

[72] Bigourd M, Dutin C F, Vanvincq O, et al. Numerical analysis of broadband fiber optical parametric amplifiers pumped by two chirped pulses. J. Opt. Soc. Am. B, 2016, 33: 1800～1807.

[73] Sun R, Jin D, Tan F, et al. High power femtosecond all-fiber chirped pulse amplification system based on Cherenkov radiation//Conference on Lasers and Electro-Optics, OSA Technical Digest (2016) (Optical Society of America, 2016), paper STu1P.7.

[74] Chen Y, Wang Y Z, Wang L J, et al. High dispersive mirrors for erbium-doped fiber chirped pulse amplification system. Opt. Express, 2016, 24: 19835～19840.

[75] Ionov P I, Rose T S. SBS reduction in nanosecond fiber amplifiers by frequency chirping. Opt. Express, 2016, 24: 13763～13777.

[76] Sanchez A, Hemmer M, Baudisch M, et al. 7μm, ultrafast, sub-millijoule-level mid-infrared optical parametric chirped pulse amplifier pumped at 2μm. Optica, 2016, 3: 147～150.

[77] Bigourd D, Dutin C F, Vanvincq O, et al. Numerical analysis of broadband fiber optical parametric amplifiers pumped by two chirped pulses. J. Opt. Soc. Am. B, 2016, 33: 1800～1807.

[78] Tan F Z, Shi H X, Sun R Y, et al. 1 μJ, sub-300 fs pulse generation from a compact thulium-doped chirped pulse amplifier seeded by Raman shifted erbium-doped fiber laser. Opt. Express, 2016, 24: 22461～22468.

[79] Hasegawa A. Transmission of stationary nonlinear optical pulses in dispersive dielectric fibers.

Ⅰ. Anomalous dispersion. Appl. Phys. Lett., 1973, 23(3): 142～144.
[80] 阿戈沃. 非线性光纤光学原理及应用.贾东方, 余震虹, 谈斌, 译. 北京: 电子工业出版社, 2002.
[81] 刘颂豪, 赫光生. 强光光学及其应用. 广州: 广东科技出版社, 1995.
[82] 庞小峰. 孤子物理学. 北京: 科学出版社, 1987.
[83] 杨祥林, 温扬敬. 光纤孤子通信理论基础. 北京: 国防工业出版社, 2000.
[84] 黄景宁, 徐济仲, 熊吟涛. 孤子概念、原理和应用. 北京: 高等教育出版社, 2004.
[85] Ablowitz M J, Kaup D J, Newell A C, et al. The inverse scattering transform-fourier analysis for nonlinear problems. Studies in Applied Mathematics, 1974, 53(4): 249～315.
[86] Mollenauer L F, Stolen R H, Gordon J P. Experimental observation of picosecond pulse narrowing and solitons in optical fiber. Phys. Rev. Lett., 1980, 45(13): 1095～1098.
[87] Mollenauer L F, Lichtman E, Harvey G T, et al. Demonstration of error-free soliton transmission over more than 15000km at 5Gbit/s single-channel, and over 11000km at l0Gbit/s in two-channel WDM. Electronics Letters, 1992, 28(8): 792～794.
[88] Favre F, Guen D L, Devaux F. 4×20Gbit/s soliton WDM transmission over 2000 km with 100 km dispersion compensated spans of standard fibre. Electronics Letters, 1997, 33 (14): 1234～1235.
[89] Favre F, Guen D L, Moulinard M L, et al. 16×20Gbit/s soliton WDM transmission over 1300 km with 100 km dispersion compensated spans of standard fibre. Electronics Letters, 1997, 33 (25): 2135～2136.
[90] Sahara A, Kubota H, Nakazawa M. Experiment and analyses of 20-Gbit/s soliton transmission systems using installed optical fiber cable. Electronics and Communications in Japan, Part 2, 1998, 81(1): 74～83.
[91] Nakazawa M, Suzuki K, Kubota H. 160Gbit/s (80Gb/s×2channels) WDM soliton transmission over 10000km using in-line synchronous modulation. Electronics Letters, 1999, 35(16): 1358～1359.
[92] Takushima Y C, Douke T, Wang X M, et al. Dispersion tolerance and transmission distance 1000km of a 40-Gb/s dispersion management soliton transmission system. Journal of Lightwave Technology, 2002, 20(3): 360～367.
[93] Gouveia-Neto A S, Wigley P G J, Taylor J R. Soliton generation through Raman amplification of noise bursts. Optics Letters, 1989, 14(20): 1122～1124.
[94] Iwatsuki K, Suzuki K I, Nishi S. Adiabatic soliton compression of gain-switched DFB-LD pulse by distributed fiber Raman amplification. IEEE Transactions Photonics Technology Letters, 1991, 3(12): 1074～1076.
[95] Murphy T E. 10-GHz 1.3-ps pulse generation using chirped soliton compression in a Raman gain medium. IEEE Photonics Technology Letters, 2002, 14(10): 1424～1426.
[96] Mollenauer L F, Stolen R H, Islam M N. Experimental demonstration of soliton propagation in long fibers: Loss compensated by Raman gain. Optics Letters, 1985, 10(5): 229～231.
[97] Iwatsuki K, Nishi S, Saruwatari M, et al. 5Gb/s optical soliton transmission experiment using Raman amplification for fiber-loss compensation. IEEE Photonics Technology Letters, 1990,

2(7): 507～509.

[98] Okhrimchuk A G, Onishchukov G, Lederer F. Long-haul soliton transmission at 1.3 m using distributed Raman amplification. Journal of Lightwave Technology, 2001, 19(6): 837～841.

[99] Ereifej H N, Grigoryan V, Carter G M. 40 Gbit/s long-haul transmission in dispersion-managed soliton system using Raman amplification. Electronics Letters, 2001, 37(25): 1538～1539.

[100] Pincemin E, Hamoir D, Audouin O, et al. Distributed-Raman-amplification effect on pulse interactions and collisions in long-haul dispersion -managed soliton transmissions. J. Opt. Soc. Am. B, 2002, 19(5): 973～980.

[101] Tio A A B, Shum P. Propagation of optical soliton in a fiber Raman amplifier. Proceedings of SPIE, 2004, 5280: 676～681.

[102] Mollenauer L F, Smith K. Demonstration of soliton transmission over more than 4000 km in fiber with loss periodically compensated by Raman gain. Optics Letters, 1988, 13(8): 675～677.

[103] Chi S, Wen S. Interaction of optical solitons with a forward Raman pump wave. Optics Letters, 1989, 14(1): 84～86.

[104] Wen S, Wang T Y, Chi S. The optical soliton transmission amplified by bidirectional Raman pumps with nonconstant depletion. IEEE Journal of Quantum Electronics, 1991, 21(8): 2066～2073.

[105] Levy G F. Raman amplification of solitons in a fiber optic ring. Journal of Lightwave Technology, 1996, 14(1): 72～76.

[106] Jubgang D Jr. F, Dikandé A M. Pulse train uniformity and nonlinear dynamics of soliton crystals in mode-locked fiber ring lasers. J. Opt. Soc. Am. B, 2017, 34: 66～75.

[107] Liu M, Luo A P, Luo Z C, et al. Dynamic trapping of a polarization rotation vector soliton in a fiber laser. Opt. Lett., 2017, 42: 330～333.

[108] Rasskazov G, Ryabtsev A, Charan K, et al. Numerical simulations of fast-axis instability of vector solitons in mode-locked fiber lasers. Opt. Express, 2017, 25: 1131～1141.

[109] Liu M, Luo A P, Yan Y R, et al. Successive soliton explosions in an ultrafast fiber laser. Opt. Lett., 2016, 41: 1181～1184.

[110] Braud F, Conforti M, Cassez A, et al. Transformation of a dispersive wave into a fundamental soliton//Conference on Lasers and Electro-Optics, OSA Technical Digest (2016) (Optical Society of America, 2016), paper STh3O.6.

[111] Wan A, Schibli T R, Li P, et al. Intensity Noise Coupling in Soliton Fiber Oscillators// Conference on Lasers and Electro-Optics, OSA Technical Digest (2016) (Optical Society of America, 2016), paper STh3O.2.

[112] Tang Y, Wright L, Charan K, et al. Intense Mid-Infrared Few-Cycle Soliton Generation Covering 2-4.3μm in Fluoride Fiber//Photonics and Fiber Technology 2016 (ACOFT, BGPP, NP), OSA Technical Digest (online) (Optical Society of America, 2016), paper AW2B.1.

[113] Bao C J, Liao P C, Zhang L, et al. Effect of a breather soliton in Kerr frequency combs on optical communication systems. Opt. Lett., 2016, 41: 1764～1767.

[114] Luo J Q, Sun B, Ji J H, et al. High-efficiency femtosecond Raman soliton generation with a tunable wavelength beyond 2 μm. Opt. Lett., 2017, 42: 1568～1571.

[115] Díaz-Otero F J, Guillán-Lorenzo O, Pedrosa-Rodriguez L. Higher-order interaction-induced effects between strong dispersion-managed solitons with dissimilar powers. J. Opt. Soc. Am. B, 2017, 34: 12～17.

[116] Li Q, Cheng Z H. Cascaded photonic crystal fibers for three-stage soliton compression. Appl. Opt. , 2016, 55: 8868～8875.

[117] Maruta A. Solitons and Nonlinear Fourier Transformation//Optical Fiber Communication Conference, OSA Technical Digest (online) (Optical Society of America, 2017), paper Th3J.3.

[118] Pickartz S, Bandelow U, Amiranashvili S. Asymptotically stable compensation of the soliton self-frequency shift. Opt. Lett., 2017, 42: 1416～1419.

[119] Nicholson J W, DeSantolo A, Zach A, et al. Soliton self-frequency shifting in a polarization-maintaining, Erbium-doped, very-large-mode-area fiber amplifier//Conference on Lasers and Electro-Optics, OSA Technical Digest(2016) (Optical Society of America, 2016), paper STh3O.4.

[120] Rishoj L, Prabhakar G, Demas J, et al. 30 nJ, ～50 fs all-fiber source at 1300 nm using soliton shifting in LMA HOM fiber //Conference on Lasers and Electro-Optics, OSA Technical Digest (2016) (Optical Society of America, 2016), paper STh3O.3.

[121] Milián C, Marest T, Kudlinski A, et al. Spectral wings of the fiber supercontinuum and the dark-bright soliton interaction. Opt. Express, 2017: 25: 10494～10499.

[122] Chen K, Wu T, Wei H Y, et al. Quantitative chemical imaging with background-free multiplex coherent anti-Stokes Raman scattering by dual-soliton Stokes pulses. Biomed. Opt. Express, 2016, 7: 3927～3939.

[123] Chen K, Wu T, Wei H, et al. Background-free coherent anti-stokes Raman spectroscopy by all-fiber-generated dual-soliton as Stokes pulse //Conference on Lasers and Electro-Optics, OSA Technical Digest (2016) (Optical Society of America, 2016), paper SF1O.3.

[124] Dantus M. Characterization and adaptive compression of a multi-soliton laser source. Opt. Express, 2017, 25: 320～329.

[125] Rasskazov G, Ryabtsev A, Charan K, et al. Multi-soliton pulse characterization and compression //International Conference on Ultrafast Phenomena, OSA Technical Digest (online) (Optical Society of America, 2016), paper UTh4A.27.

[126] Jossent M, Cadroas P, Kotov L V, et al. Single mode Soliton in few-mode fiber //Advanced Photonics 2016 (IPR, NOMA, Sensors, Networks, SPPCom, SOF), OSA Technical Digest (online) (Optical Society of America, 2016), paper SoM4G.2.

[127] Buch S, Agrawal G P. Intermodal soliton interaction in nearly degenerate modes of a multimode fiber. J. Opt. Soc. Am. B, 2016, 33: 2217～2224.

[128] Chekhovskoy, Rubenchik A, Shtyrina O V, et al. Nonlinear pulse combining and compression in multi-core fibers with hexagonal lattice //Photonics and Fiber Technology 2016 (ACOFT, BGPP, NP), OSA Technical Digest (online) (Optical Society of America, 2016), paper NTh4A.5.

[129] Liu S L, Wang W Z, Xu J Z. Exact N-soliton solutions of the modified nonlinear Schrödinger equation. Phys. Rev. E, 1994, 49 (6): 5726～5730.

[130] Liu S L, Wang W Z. Complete compensation for the soliton self-frequency shift and third-order dispersion of a fiber. Opt. Lett., 1993,18(22): 1911～1912.

[131] Liu S L, Liu X Q. Mutual compensation of the higher-order nonlinearity and the third-order dispersion. Phys. Lett. A, 1997, 225(1-3): 67～72.

[132] 高以智, 姚敏玉, 许宝西, 等. 2.5GHz 光孤子传输. 高技术通讯, 1994, 7: 4～6.

[133] 许宝西, 李京辉, 姜新, 等. 2.5GHz 光孤子传输. 电子学报, 1995, 23(11): 38～54.

[134] 杨祥林, 毛庆和, 温扬敬, 等. 30km 2.5GHz 光孤子波传输与压缩实验研究. 高技术通讯, 1996,10:26～28.

[135] 余建军, 杨伯君, 余建国, 等. 光孤子传输实验研究. 光电子 · 激光, 1996, 7(5): 267～272.

[136] 余建军, 杨伯君, 管克俭. 5GHz 的 16.2ps 超短光脉冲的产生. 光学学报, 1998, 18(1) :14～17.

[137] 余建军, 杨伯君, 管克俭. 基于不同色散光纤的光纤链的孤子传输研究. 光学学报, 1998, 18(4): 446～450.

[138] 张晓光, 林宁, 张涛, 等. 预啁啾 10GHz, 38km 色散管理孤子的传输实验. 光子学报, 2001, 30(7): 813～817.

[139] 曹文华, 刘颂豪, 廖常俊, 等. 色散缓变光纤中的孤子效应拉曼脉冲产生. 中国激光, 1994, 21(6): 489～494.

[140] 李宏, 杨祥林, 刘堂坤. 暗孤子传输系统中调制拉曼泵浦的控制作用. 中国激光, 1997, 24(7): 654～658.

[141] 沈廷根, 郑浩, 李正华, 等. 掺杂光子晶体光纤的缺陷模增益谱与光孤子拉曼放大研究. 人工晶体学报, 2005, 34(6): 1065～1073.

[142] 唐慧斌, 周文龙, 宋丽军. 锁模光纤激光器中矢量孤子的传输特性研究. 量子光学学报, 2014, 20(4): 311～315.

[143] 霍佳雨, 郭玉彬, 王珂. 被动锁模掺镱光纤激光器中耗散孤子特性. 北京邮电大学学报, 2016, 39(2): 30～31.

[144] 张解放, 赵辟, 胡文成, 等. 非均匀非线性波导中涡旋光孤子的相互作用传播. 光学学报, 2013, 33(4): 0419001-1～8.

[145] 王丽, 杨荣草, 贾鹤萍, 等. 色散渐减光纤链中孤子的周期集总放大和恢复. 光学学报, 2017, (6): 0619001.

[146] 武达, 王娟芬, 石佳, 等. 掺杂光纤中 Peregrine 孤子的产生和传输. 光学学报, 2017, 37(04): 0406002.

[147] 张露, 张健, 肖燕. 啁啾艾里脉冲和孤子在非均匀光纤中的传输.中国激光, 2017, 44(09): 0906005.

[148] 时雷, 马挺, 吴浩煜, 等. 基于耗散孤子种子的啁啾脉冲光纤放大系统输出特性. 物理学报, 2016, 65(8): 084203.

[149] 杨建菊, 周桂耀, 韩颖, 等. 基于光子晶体光纤和飞秒激光源的近红外波段宽带孤子和可见区高效色散波产生的实验. 红外与毫米波学报, 2016, 35(4): 477～482.

[150] Matos C J S, Talor J R. Tunable repetition-rate multiplication of a 10 GHz pulse train using linear and nonlinear fiber propagation. Applied Physics Letters, 2003, 83(26): 5356～5358.

[151] 王安斌, 伍剑, 拱伟, 等. 高消光比超短脉冲产生的实验研究. 中国激光, 2004, 31(3):

265～268.

[152] 王兴涛, 印定军, 帅斌, 等. 应用全反射二阶自相关仪测量超短脉冲脉宽. 中国激光, 2004, 31(8): 1018～1020.

[153] Lin Q, Wright K, Agrawal G P, et al. Spectral responsitivity and efficiency of metal-based femtosecond autocorrelation technique. Optics Communications, 2004, 242: 279～283.

[154] Dai J M, Teng H, Guo C L. Second- and third-order interferometric autocorrelations based on harmonic generations from metal surfaces. Optics Communications, 2005, 252: 173～178.

[155] Fittinghoff D N, Au J A D, Squier J. Spatial and temporal characterizations of femtosecond pulses at high-numerical aperture using collinear, background-free, third-harmonic autocorrelation. Optics Communications, 2005, 247: 405～426.

[156] Wang S, Wang Y B, Feng G Y, et al. Generation of double-scale pulses in a LD-pumped Yb:phosphate solid-state laser. Appl. Opt., 2017, 56: 897～900.

[157] Le M A, Guilbaud O, Larroche O, et al. Evidence of partial temporal coherence effects in the linear autocorrelation of extreme ultraviolet laser pulses. Opt. Lett., 2016, 41: 3387～3390.

[158] Chaparro A, Furfaro L, Balle S. Subpicosecond pulses in a self-starting mode-locked semiconductor-based figure-of-eight fiber laser. Photon. Res., 2017, 5: 37～40.

[159] Lauterio-Cruz P, Hernandez-Garcia J C, Pottiez O, et al. High energy noise-like pulsing in a double-clad Er/Yb figure-of-eight fiber laser. Opt. Express, 2016, 24: 13778～13787.

[160] Lin J H, Chen C L, Chan C W, et al. Investigation of noise-like pulses from a net normal Yb-doped fiber laser based on a nonlinear polarization rotation mechanism. Opt. Lett., 2016, 41: 5310～5313.

[161] Zhang F, Fan X W, Liu J, et al. Dual-wavelength mode-locked operation on a novel Nd^{3+},Gd^{3+}: SrF_2crystal laser. Opt. Mater. Express, 2016, 6: 1513～1519.

[162] Chao M S, Cheng H N, Fong B J, et al. High-sensitivity ultrashort mid-infrared pulse characterization by modified interferometric field autocorrelation. Opt. Lett., 2015, 40: 902～905.

[163] Sun B, Salter P S, Booth M J. Pulse front adaptive optics: a new method for control of ultrashort laser pulses. Opt. Express, 2015, 23: 19348～19357.

[164] Lin S S, Hwang S K, Liu J M. High-power noise-like pulse generation using a 1.56-μm all-fiber laser system. Opt. Express, 2015, 23: 18256～18268.

[165] Traore A, Lalanne E, Johnson A M. Determination of the nonlinear refractive index of multimode silica fiber with a dual-line ultra-short pulse laser source by using the induced grating autocorrelation technique. Opt. Express, 2015, 23: 17127～17137.

[166] Suzuki M, Ganeev R A, Yoneya S, et al. Generation of broadband noise-like pulse from Yb-doped fiber laser ring cavity. Opt. Lett., 2015, 40: 804～807.

[167] Tian W L, Wang Z H, Liu J X, et al. Dissipative soliton and synchronously dual-wavelength mode-locking Yb:YSO lasers. Opt. Express, 2015, 23: 8731～8739.

[168] Agrawal A P. Nonlinear Fiber Optics. 5th edition. Singapore: Elsevier Inc. Elsevier Pte Ltd. 2012.

[169] Schelev M Y, Richardson M C, Alcock A J. Image-converter streak camera with picosecond resolution. Appl. Phys. Lett., 1971, 18(8): 354～357.

[170] Kane D J, Trebino R. Characterization of arbitrary femtosecond pulses using frequency-resolved optical gating. IEEE J. Quantum Electron., 1993, 29(2): 571～579.

[171] DeLong K W, Trebino R, Hunter J, et al. Frequency-resolved optical gating with the use of second-harmonic generation. J. Opt. Soc. Am. B,1994, 11(11): 2206～2215.

[172] Gallmann L, Steinmeyer G, Sutter D H, et al. Collinear type II second-harmonic -generation frequency-resolved optical gating for the characterization of sub-10-fs optical pulses. Optics Letters, 2000, 25(4): 269～271.

[173] 王兆华, 魏志义, 滕浩, 等. 飞秒激光脉冲的谐波频率分辨光学开关法测量研究. 物理学报, 2003, 52(2): 362～366.

[174] 龙井华, 高继华, 巨养锋, 等. 用SHG-FROG方法测量超短光脉冲的振幅和相位. 光子学报, 2002, 31(10): 1292～1296.

[175] DeLong K W, Fittinghoff D N, Trebino R, et al. Pulse retrieval in frequency-resolved optical gating based on the method of generalized projections. Optics Letters, 1994, 19(24): 2152～2154.

[176] Hu J, Zhang G Z, Zhang B G, et al. Using frequency-resolved optical gating to retrieve amplitude and phase of ultrashort laser pulse. Journal of Optoeletronics · Laser, 2002, 13(3): 232～236.

[177] Lacourt P A, Dudley J M, Merolla J M, et al. Milliwatt-peak-power pulse characterization at 1.55 um by wavelength-conversion frequency-resolved optical gating. Optics Letters, 2002, 27(10): 863～865.

[178] Barry L P, Delburgo S, Thomsen B C, et al. Optimization of optical data transmitters for 40-Gb/s lightwave systems using frequecy resolved optical gating. Photon. Tech. Lett., 2002, 14(7): 971～973.

[179] Liu S, Lu D, Zhao L, et al. SHG-FROG characterization of a novel multichannel synchronized AWG-based mode-locked laser //Conference on Lasers and Electro-Optics, OSA Technical Digest (online) (Optical Society of America, 2017), paper JTh2A.131.

[180] Kane D J. Improved principal components generalized projections algorithm for frequency resolved optical gating //Conference on Lasers and Electro-Optics, OSA Technical Digest (online) (Optical Society of America, 2017), paper STu3I.4.

[181] Hyyti J, Escoto E, Steinmeyer G, et al. Interferometric time-domain ptychography for ultrafast pulse characterization. Opt. Lett., 2017, 42: 2185～2188.

[182] Sidorenko, Lahav O, Avnat Z, et al. Ptychographic reconstruction algorithm for frequency-resolved optical gating: super-resolution and supreme robustness. Optica, 2016, 3: 1320～1330.

[183] Heidt A M, Spangenberg D M, Brügmann M, et al. Improved retrieval of complex supercontinuum pulses from XFROG traces using a ptychographic algorithm. Opt. Lett., 2016, 41: 4903～4906.

[184] Ermolov A, Valtna-Lukner H, Travers J, et al. Characterization of few-fs deep-UV dispersive waves by ultra-broadband transient-grating XFROG. Opt. Lett., 2016, 41: 5535～5538.

[185] Fuji T, Shirai H, Nomura Y. Self-referenced frequency-resolved optical gating capable of

carrier-envelope phase determination //Conference on Lasers and Electro-Optics, OSA Technical Digest (2016) (Optical Society of America, 2016), paper SM3I.7.

[186] Steinmeyer A. Interferometric FROG for ultrafast spectroscopy on the few-cycle scale // Conference on Lasers and Electro-Optics, OSA Technical Digest (2016) (Optical Society of America, 2016), paper STu4I.1.

[187] Okamura B, Sakakibara Y, Omoda E, et al. Experimental analysis of coherent supercontinuum generation and ultrashort pulse generation using cross-correlation frequency resolved optical gating (X-FROG). J. Opt. Soc. Am. B, 2015, 32: 400～406.

[188] Itakura R, Kumada T, Nakano M, et al. Frequency-resolved optical gating for characterization of VUV pulses using ultrafast plasma mirror switching. Opt. Express, 2015, 23: 10914～10924.

[189] Snedden E W, Walsh D A, Jamison S P. Revealing carrier-envelope phase through frequency mixing and interference in frequency resolved optical gating. Opt. Express, 2015, 23: 8507～8518.

[190] Hause A, Kraft S, Rohrmann P, et al. Reliable multiple-pulse reconstruction from second-harmonic-generation frequency-resolved optical gating spectrograms. J. Opt. Soc. Am. B, 2015, 32: 868～877.

[191] Li X J, Liao J L, Nie Y M, et al. Unambiguous demonstration of soliton evolution in slow-light silicon photonic crystal waveguides with SFG-XFROG. Opt. Express, 2015, 23: 10282～10292.

[192] Marcuse D. Pulse distortion in single-mode fibers. 3: Chirped pulses. Applied Optics, 1981, 20(20): 3573～3579.

[193] Lassen H E, Mengel F, Tromborg B, et al. Evolution of chirped pulses in nonlinear single-mode fibers. Optics Letters,1985, 10 (1): 34～36.

[194] Planas S A, Mansur N L, Cruz C H, et al. Spectral narrowing in the propagation of chirped pulses in single-mode fibers. Optics Letters, 1993, 18(9): 699～701.

[195] Cundiff S T, Collings B C, Boivin L, et al. Propagation of highly chirped pulses in fiber-optic communications systems. Journal of Lightwave Technology, 1999,17(5): 811～816.

[196] Wu H C, Sheng Z M, Zhang J. Chirped pulse compression in nonuniform plasma Bragg gratings. Applied Physics Letters, 2005, 87(20): 219.

[197] Rao M, Sun X H, Zhang M D. A modified split-step Fourier method for optical pulse propagation with polarization mode dispersion. Chinese Physics, 2003, 12(5): 502～506 .

[198] Pan L Z, Lü B D. Anomalous spectral behaviour of diffracted chirped Gaussian pulses in the near field. Chinese Physics, 2004, 13(5): 637～645.

[199] Wang J, Wang Z L. Evolution of sum-chirp in polarization multiplexed communication system. Chinese Physics, 2004, 13(6): 877～881.

[200] 舒学文，黄德修，阮玉．啁啾高斯脉冲经啁啾光纤光栅反射后的传输特性．光学学报，1999, 19(10): 1305～1309.

[201] Richardson D R, Stauffer H U, Roy S, et al. Comparison of chirped-probe-pulse and hybrid femtosecond/picosecond coherent anti-Stokes Raman scattering for combustion thermometry.

Appl. Opt., 2017, 56: E37～E49.

[202] Giree A, Mero M, Arisholm G, et al. Numerical study of spatiotemporal distortions in noncollinear optical parametric chirped-pulse amplifiers. Opt. Express, 2017, 25: 3104～3121.

[203] Jubgang D J F, Dikandé A M. Pulse train uniformity and nonlinear dynamics of soliton crystals in mode-locked fiber ring lasers. J. Opt. Soc. Am. B, 2017, 34: 66～75.

[204] Sanchez D, Hemmer M, Baudisch M, et al. 7 μm, ultrafast, sub-millijoule-level mid-infrared optical parametric chirped pulse amplifier pumped at 2 μm. Optica, 2016, 3: 147～150.

[205] Lin S F, Lin Y H, Cheng C H, et al. Stability and chirp of tightly bunched solitons from nonlinear polarization rotation mode-locked erbium-doped fiber lasers. J. Lightwave Technol., 2016, 34: 5118～5128.

[206] Klaus M, Shaw J K. Influence of pulse shape and frequency chirp on stability of optical solitons. Optics Communications, 2001, 197(4-6): 491～500.

[207] Desaix M, Helczynski L, Anderson D, et al. Propagation properties of chirped soliton pulses in optical nonlinear Kerr media. Phys. Rev. E, 2002, 65(5): 056602.

[208] Li Z H, Li L, Tian H P, et al. Chirped femtosecond solitonlike laser pulse form with self-frequency shift. Physical Review Letters, 2002, 89(26): 263901.

[209] MacLeod A M, Yan X, Gillespie W A, et al. Formation of low time-bandwidth product, single-sided exponential optical pulses in free-electron laser oscillators. Physical Review E, 2000, 62(3): 4216～4220.

[210] Shapiro S L. 超短光脉冲-皮秒技术及其应用. 朱世清, 译. 北京: 科学出版社, 1987.

[211] 朱京平. 光电子技术基础. 成都: 四川科学技术出版社, 2003.

[212] Mori K, Takara H, Kawanishi S, et al. Flatly broadened supercontinuum generation in a dispersion decreasing fiber with convex dispersion profile. Electronics Letters, 1997, 33(21): 1806～1807.

[213] Nan Y B, Lou C Y, Wang J P, et al. Improving the performance of a multiwavelength continuous-wave optical source based on supercontinuum by suppressing degenerate four-wave mixing. Optics Communications, 2005, 256: 428～434.

[214] Yang J W, Chae C J. WDM-PON upstream transmission using Fabry-Perot laser diodes externally injected by polarization-insensitive spectrum-sliced supercontinuum pulses. Optics Communications, 2006, 260: 691～695.

[215] Smirnov S V, Ania-Castanon J D, Ellingham T J, et al. Optical spectral broadening and supercontinuum generation in telecom applications. optical Fiber Technology, 2006, 12(2): 122～147.

[216] Wang H C, Alismail A, Barbiero G, et al. Cross-polarized, multi-octave supercontinuum generation. Opt. Lett., 2017, 42: 2595～2598.

[217] Petersen C R, Engelsholm R D, Markos C, et al. Increased mid-infrared supercontinuum bandwidth and average power by tapering large-mode-area chalcogenide photonic crystal fibers. Opt. Express, 2017, 25: 15336～15348.

[218] Yin K, Zhang B, Yang L Y, et al. 15.2 W spectrally flat all-fiber supercontinuum laser source with >1 W power beyond 3.8 μm. Opt. Lett., 2017, 42: 2334～2337.

[219] Valliammai M, Sivabalan S. Wide-band supercontinuum generation in mid-IR using polarization maintaining chalcogenide photonic quasi-crystal fiber. Appl. Opt., 2017, 56: 4797～4806.

[220] Khalifa A B, Salem A B, Cherif R. Mid-infrared supercontinuum generation in multimode As_2Se_3 chalcogenide photonic crystal fiber. Appl. Opt., 2017, 56: 4319～4324.

[221] Eftekhar M A, Wright L G, Mills M S, et al. Versatile supercontinuum generation in parabolic multimode optical fibers. Opt. Express, 2017, 25: 9078～9087.

[222] Zhao S L, Yang H, Zhao C J, et al. Harnessing rogue wave for supercontinuum generation in cascaded photonic crystal fiber. Opt. Express, 2017, 25: 7192～7202.

[223] Kedenburg S, Strutynski C, Kibler B, et al. High repetition rate mid-infrared supercontinuum generation from 1.3 to 5.3 μm in robust step-index tellurite fibers. J. Opt. Soc. Am. B, 2017, 34: 601～607.

[224] Strutynski C, Froidevaux P, Désévédavy F, et al. Tailoring supercontinuum generation beyond 2 μm in step-index tellurite fibers. Opt. Lett., 2017, 42: 247～250.

[225] Tarnowski K, Martynkien T, Mergo P, et al. Coherent supercontinuum generation up to 2.2 μm in an all-normal dispersion microstructured silica fiber. Opt. Express, 2016, 24: 30523～30536.

[226] 娄采云, 李玉华, 伍剑, 等. 利用10GHz主动锁模光纤激光器在DSF中产生超连续谱. 中国激光, 2000, 27(9): 814～818.

[227] 伍剑, 李玉华, 娄采云, 等. 利用超连续谱光源产生超短光脉冲.光学学报, 2000, 20(3): 325～329.

[228] 王肇颖, 贾东方, 葛春风, 等. 10 GHz 再生锁模光纤激光器获得光纤超连续谱的研究. 光电子 · 激光, 2006, 17(3): 9～13.

[229] Szkulmowski M, Wojtkowski M, Bajraszewski T, et al. Quality improvement for high resolution in vivo images by spectral domain optical coherence tomography with supercontinuum source. Optics Communications, 2005, 246: 569～578.

[230] Kano H, Hamaguchi H. Femtosecond coherent anti-Stokes Raman scattering spectroscopy using supercontinuum generated from a photonic crystal fiber. Applied Physics Letters, 2004, 85(19): 4298～4300.

[231] Lindfors K, Kalkbrenner T, Stoller P, et al. Detection and spectroscopy of gold nanoparticles using supercontinuum white light confocal microscopy. Physical Review Letters, 2004, 93(3): 037401.

[232] Mori K, Takara H, Kawanishi S. Analysis and design of supercontinuum pulse generation in a single-mode optical fiber. J. Opt. Soc. Am. B, 2001, 18(12): 1780～1792.

[233] 娄采云, 高以智, 王建萍, 等. 光纤中超连续谱产生的理论与实验研究. 清华大学学报(自然科学版), 2003, 43(4): 441～445.

[234] 贾东方, 丁永奎, 胡志勇, 等. 光纤中超连续谱产生机理研究. 光电子 · 激光, 2004, 15(5): 612～620.

[235] 王肇颖, 王永强, 李智勇, 等. 皮秒脉冲在色散位移光纤中产生的超连续谱. 光电子 · 激光, 2004, 15(5): 528～533.

[236] 李智勇, 王肇颖, 王永强, 等. 基于 100m色散位移光纤的超连续谱实验研究. 光子学报,

2004, 33(9): 1064～1066.

[237] 成纯富, 王晓方, 鲁波. 飞秒光脉冲在光子晶体光纤中的非线性传输和超连续谱产生. 物理学报, 2004, 53(6): 1826.

[238] 陈泳竹, 李玉忠, 屈圭, 等. 反常色散平坦光纤产生平坦宽带超连续谱的数值研究. 物理学报, 2006, 55(2): 0717.

[239] Stolen R H, Ashikin A. Optical Kerr effect in glass waveguide. Appl. Phys. Lett., 1973, 22(6): 294～296.

[240] Kitayama K I, Kimura Y, Okamoto K, et al. Optical sampling using an all-fiber optical Kerr shutter. Appl. Phys. Lett., 1985, 46(7): 623～625.

[241] Islam M N, Soccolich C E, Chen C J, et al. All-optical inverter with one picojoule switching energy. Electronics Letters, 1991, 27(2): 130～132.

[242] Asobe M, Kanamori T, Kubodera K. Ultrafast all-optical switching using highly nonlinear chalcogenide glass fiber. IEEE Photon. Technol. Lett., 1992, 4(4): 362～365.

[243] Islam M N, Sauer J R. GEO modules as a natural basis for all-optical fiber logic systems. IEEE J. Quantum Electron., 1991, 27(3): 843～848.

[244] Menyuk C R. Nonlinear pulse propagation in birefringent optical fibers. IEEE J. of Quantum Electron., 1987, 23(2): 174～176.

[245] Menyuk C R. Stability of solitons in birefringent optical fibers. II. Arbitrary amplitudes. J. Opt. Soc. Am. B, 1988, 5(2): 392～402.

[246] Menyuk C R. Stability of solitons in birefringent optical fibers. I: Equal propagation amplitudes. Optics Letters, 1987, 12(8): 614～616.

[247] Cao X D, Meyerhofer D D. Soliton collisions in optical birefringent fibers. J. Opt. Soc. Am. B, 1994, 11(2): 380～385.

[248] 唐雄燕, 叶培大. 双折射光纤中正交极化孤子碰撞的数值研究. 北京邮电大学学报, 1994, 17(2): 10～17.

[249] 江辉, 庞勇, 蒋佩璇. 双折射光纤中孤子碰撞的数值研究. 北京邮电大学学报, 1996, 19(4): 42～47.

[250] 黄洪涛, 聂再清. 线双折射光纤与正交极化孤子碰撞的研究. 中国激光, 1999, 26(6): 163～170.

[251] Zhao N B, Li X Y, Li G F, et al. Capacity limits of spatially multiplexed free-space communication. Nature Photonics, 2015, 9: 822～826.

[252] Urick V J, Bucholtz F, McKinney J D, et al. Long-haul analog photonics. J. Lightwave Technol., 2011, 29: 1182～1205.

[253] Wen H, Zheng H J, Zhu B Y, et al. Experimental demonstration of long-distance analog transmission over few-mode fibers. Optical Fiber Communications Conference & Exhibition, 2015.

[254] Wen H, Zheng H j, Mo Q, et al. Analog fiber-optic links using high-order fiber modes. European Conference on Optical Communication, 2015.

[255] Effenberger F J. Space division multiplexing in access networks. Proc. SPIE, 2015, 9387: 938704-1～938704-6.

[256] Cen X, Chand N, Velazquez-Benitez A M, et al. Demonstration of world's first few-mode GPON. Proc.Eur. Conf. Opt. Commun., 2014: 1～3.

[257] Xia C, Chand N, Velázquez-Benítez A M, et al. Time-division-multiplexed few-mode passive optical network. Opt. Exp., 2015, 23: 1151～1158.

[258] Li B, Feng Z, Tang M, et al. Experimental demonstration of large capacity WSDM optical access network with multicore fibers and advanced modulation formats. Opt. Exp., 2015, 23(9): 10997～11006.

[259] Ren F, Li J, Hu T, et al. Cascaded mode-division-multiplexing and time-division-multiplexing passive optical network based on low mode-crosstalk FMF and mode MUX/DEMUX. IEEE Photon. J., 2015, 7(5): 1～9.

[260] Wen H, Xia C, Velazquez-Benitez A, et al. First Demonstration of 6-mode PON achieving a record gain of 4 dB in upstream transmission loss budget. Journal of Lightwave Technology, 2016, 34(8): 1990～1996.

[261] Xia C, Wen H, Velázquez-Benítez A M, et al. Experimental demonstration of 5-mode PON achieving a net gain of 4 dB in upstream transmission loss budget. European Conference on Optical Communication, 2015: 1～3.

第 2 章　啁啾脉冲实验测量原理与啁啾脉冲自相关特性

近几年，频率分辨光学门(FROG)新型测量技术受到了国际科学研究工作者的普遍关注[1-23]，脉冲自相关频谱和强度自相关参量是 FROG 测量技术的重要参量，是其进行脉冲测量时的监视窗口。研究这些参量的变化，不仅可以较早地预测被测脉冲的特性、提高测量效率，而且可以指导人们准确测量脉冲特性。超高斯脉冲作为重要的脉冲源，广泛应用于不同的研究领域。在强激光技术领域存在许多强度分布近似为平顶的光束，可用超高斯光束描述[24, 25]，例如，聚变激光驱动器中的强激光，超高斯反射率镜腔输出的激光等。在实际的光通信系统中，人们通常采用由直接调制半导体激光器产生的脉冲作为信号脉冲，实验表明这种脉冲的波形具有超高斯分布，并且通常还带有较大的啁啾[1, 26, 27]。然而，超高斯脉冲的自相关频谱和强度自相关研究罕见报道，因此对超高斯脉冲自相关特性的研究就显得非常重要。

考虑到实际脉冲源通常为具有较大频率啁啾且频率啁啾可以通过改变传输长度、采用啁啾光栅技术或预啁啾技术等进行调节[1]，本章研究了啁啾脉冲自相关特性及其受锐度参量 m、啁啾参量(chirp parameter)C、脉冲噪声和随机噪声影响的变化规律，与原脉冲相应参量作了比较，给出了一个有效滤除随机噪声的方法并作了实验验证，为我们使用二次谐波-频率分辨光学门(SHG-FROG)脉冲分析仪或者基于自相关技术原理的仪器测量脉冲特性提供了重要参考。

2.1　频率分辨光学门技术原理

2.1.1　二次谐波-频率分辨光学门脉冲分析仪的数据测量

图 2-1 所示是 SHG-FROG 脉冲分析仪示意图。由图 2-1 可见，它与传统的强度自相关测量仪类似，只要把光谱仪(spectrometer)去掉，只留下 CCD(电荷耦合器件)，或者换成一般的能量探测器件(相对脉冲是慢探测器)，就成了传统的无背景强度自相关测量仪。待测脉冲 $E(T)$ 首先被分束器(splitter)分为两束，其中一束作为探测(probe)脉冲，另一束作为快门开关(gate)脉冲 $E(T-\tau')$，由步进电机控制的

平移台为快门开关光束提供时间延迟 τ'，用凸透镜(len)将两束光会聚在 $LiNbO_3$ 的倍频晶体(crystal)中心。探测脉冲与快门开关脉冲通过在倍频晶体中相互作用产生自相关脉冲

$$E_s(T,\tau') = AE(T)E(T-\tau') \tag{2-1}$$

式中，A 为与倍频晶体有关的比例系数。

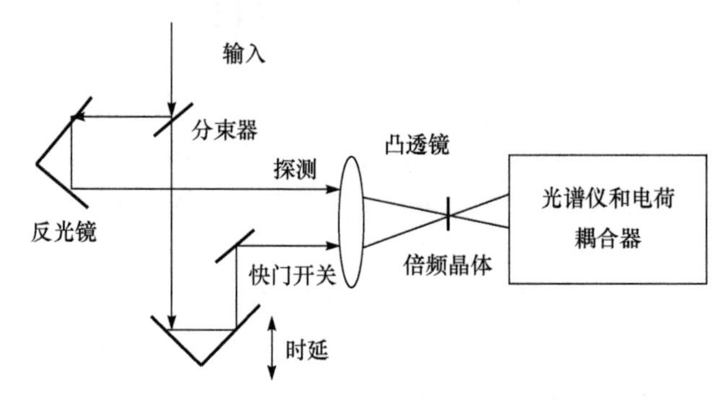

图 2-1　SHG-FROG 脉冲分析仪示意图

强度自相关测量仪是测量自相关脉冲强度

$$G_1(\tau') = \int_{-\infty}^{\infty} \left|E_s(T,\tau')\right|^2 \mathrm{d}t = A^2 \int_{-\infty}^{\infty} \left|E(T)\right|^2 \left|E(T-\tau')\right|^2 \mathrm{d}T \tag{2-2}$$

$G_1(\tau')$ 称为强度自相关函数，其图形称为自相关曲线。自相关曲线并不是待测脉冲$|E(T)|^2$ 的波形曲线，而是自相关脉冲的波形曲线。自相关曲线不携带待测脉冲的形状信息，其半峰全宽(FWHM) $\Delta\tau'_G$ 和待测脉冲半峰全宽 ΔT 满足关系 $\Delta T = \Delta\tau'_G/k$,其中比例系数 k 与待测脉冲的形状有关。例如，对高斯脉冲 k =1.414，对双曲正割脉冲 k =1.55。若已知待测脉冲的波形，根据自相关曲线的宽度便可推知待测脉冲的宽度。在实际测量中，待测脉冲形状往往是未知的，难以选择比例系数 k。因此，用自相关测量仪有很大的局限性。

SHG-FROG 脉冲分析仪是在每一个延迟位置 $-N\tau'_0$，$-(N-1)\tau'_0$，…，$-\tau'_0$, 0, τ'_0, …, $(N-1)\tau'_0$, $N\tau'_0$ 测量自相关脉冲的频谱 $\tilde{E}_s(\omega,\tau')$，这里 τ'_0 是最小时间延迟，N 是正整数。自相关脉冲的频谱经光谱仪展开后，用 CCD 测量光谱强度，得到一个与时延和频率有关的二维函数

$$I_{\text{test}}(\omega,\tau') = \left|\tilde{E}_{\text{test}}(\omega,\tau')\right|^2 = \left|\int_{-\infty}^{\infty} E_s(T,\tau')\exp(\mathrm{i}\omega T)\mathrm{d}T\right|^2 \tag{2-3}$$

该函数的图形称为 FROG 图。对 FROG 图数据进行迭代运算即可恢复得到脉冲的脉宽、谱宽、波形、相位和啁啾等特征信息。

2.1.2　强度自相关曲线和自相关频谱曲线

将式(2-3)对频率ω积分，得脉冲的强度自相关函数

$$\begin{aligned} G(\tau') &= \int_{-\infty}^{\infty} I_{\text{test}}(\omega,\tau')\mathrm{d}\omega = \int_{-\infty}^{\infty}\left|\int_{-\infty}^{\infty} E_{\mathrm{s}}(T,\tau')\exp(\mathrm{i}\omega T)\mathrm{d}T\right|^2 \mathrm{d}\omega \\ &= A^2\int_{-\infty}^{\infty}\left|E(T)\right|^2\left|E(T-\tau')\right|^2\mathrm{d}T \end{aligned} \tag{2-4}$$

比较式(2-2)和式(2-4)可知，由所测量的 FROG 图数据$I_{\text{test}}(\omega,\tau')$给出的强度自相关函数$G(\tau')$与由强度自相关测量仪给出的强度自相关函数$G_1(\tau')$完全相同，其图形称为强度自相关曲线。同理，将式(2-3)对时延τ'积分，得

$$\begin{aligned} \int_{-\infty}^{\infty} I_{\text{test}}(\omega,\tau')\mathrm{d}\tau' &= \int_{-\infty}^{\infty}\left|\tilde{E}_{\text{test}}(\omega,\tau')\right|^2\mathrm{d}\tau' \\ &= \int_{-\infty}^{\infty}\left|\tilde{E}(\omega')\right|^2\left|\tilde{E}(\omega'-\omega)\right|^2\mathrm{d}\omega' = G(\omega) \end{aligned} \tag{2-5}$$

$G(\omega)$称为自相关频谱函数，其图形称为自相关频谱曲线。

2.1.3　波形和相位恢复算法的实现

待测脉冲电场$E(T)$的恢复是依据测量得到的 FROG 图数据$I_{\text{test}}(\omega,\tau')$重建信号的过程，属于二维相位恢复问题，具有唯一解。目前，依据$I_{\text{test}}(\omega,\tau')$重建信号的方法已经形成了不同的迭代傅里叶算法方案,典型的算法是广义投影算法。广义投影算法示意图如图 2-2 所示。先假定一个初始猜测脉冲电场$E(T)$(如高斯脉冲)，再用式(2-1)作为时域限制得到自相关脉冲信号电场$E_{\mathrm{s}}(T,\tau')$，并给出$E_{\mathrm{s}}(T,\tau')$的傅里叶变换$\tilde{E}_{\mathrm{s}}(\omega,\tau')$。将测量得到的自相关脉冲信号电场的频谱$\tilde{E}_{\text{test}}(\omega,\tau')$的幅度

$$\left|\tilde{E}_{\text{test}}(\omega,\tau')\right| = \sqrt{I_{\text{test}}(\omega,\tau')} \tag{2-6}$$

作为频域限制替换$\tilde{E}_{\mathrm{s}}(\omega,\tau')$的幅度，得

$$\tilde{E}_{\mathrm{s}}'(\omega,\tau') = \frac{\tilde{E}_{\mathrm{s}}(\omega,\tau')}{\left|\tilde{E}_{\mathrm{s}}(\omega,\tau')\right|}\cdot\sqrt{I_{\text{test}}(\omega,\tau')} \tag{2-7}$$

由$\tilde{E}_{\mathrm{s}}'(\omega,\tau')$傅里叶逆变换便得到第一次迭代后的时域自相关脉冲信号电场$E_{\mathrm{s}}'(T,\tau')$。

为了完成第二次迭代，找到新的迭代值$E_{\mathrm{s}}''(T,\tau')$，需要应用时域限制，并使$E_{\mathrm{s}}''(T,\tau')$点和$E_{\mathrm{s}}'(T,\tau')$点之间的距离最小。

$$L = \sum_{T,\tau'=1}^{N}\left|E_{\mathrm{s}}'(T,\tau') - E_{\mathrm{s}}''(T,\tau')\right|^2 = \sum_{T,\tau'=1}^{N}\left|E_{\mathrm{s}}'(T,\tau') - E'(T)E'(T-\tau')\right|^2 \tag{2-8}$$

为了实现L的最小化，令L相对$E'(T)$的导数为零，并求解得到$E'(T)$。再以此值

$E'(T)$ 作为新的脉冲电场重复上述变换，直至 FROG 误差接近收敛的标准，或者限定迭代运算次数，直至完成指定次数(作者在实际测量中采用限定迭代运算次数的方法)。

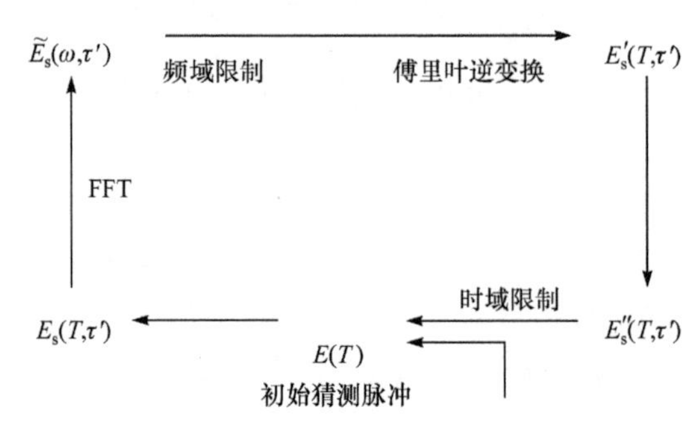

图 2-2 脉冲重建的广义投影算法示意图

2.2 啁啾脉冲自相关特性及其受噪声的影响

由频率分辨光学门技术原理式(2-1)～(2-5)可以得到脉冲的强度自相关和自相关频谱公式，并可求得脉冲的强度自相关半峰全宽、自相关频谱半峰全宽(spectral FWHM)，和脉冲的自相关时间带宽积。A 为与倍频晶体有关的比例常数，为方便计算，本书令 A=1。

对啁啾超高斯脉冲，归一化包络电场可表示为

$$U(T)=\exp\left[-\frac{1+\mathrm{i}C}{2}\left(\frac{T}{T_0}\right)^{2m}\right],\quad -\infty<T<\infty \tag{2-9}$$

式中，T 是时间，T_0 是在 e^{-1} 强度处的脉冲半宽度，C 是啁啾参量，m 为脉冲前后沿的锐度参量。m=1，对应啁啾高斯脉冲情况；m 越大，脉冲前后沿锐度越大，脉冲越接近方形脉冲或平顶脉冲。由式(2-9)得脉冲半峰全宽为

$$\Delta T=2\sqrt[2m]{\ln 2}\,T_0 \tag{2-10}$$

高斯脉冲的脉冲半峰全宽 $\Delta T_{m=1}=2\sqrt{\ln 2}\,T_0=1.6651T_0$ 是式(2-10)在 m=1 时的特例。

式(2-9)所示啁啾超高斯脉冲的频谱

$$\begin{aligned}U(\omega)&=\int_{-\infty}^{\infty}U(T)\exp(\mathrm{i}\omega T)\mathrm{d}T\\&=\int_{-\infty}^{\infty}\exp\left[-\frac{1+\mathrm{i}C}{2}\left(\frac{T}{T_0}\right)^{2m}\right]\exp(\mathrm{i}\omega T)\mathrm{d}T\end{aligned} \tag{2-11}$$

在 m=1 时(即高斯脉冲)，脉冲频谱为

$$
\begin{aligned}
U(\omega)_{m=1} &= \int_{-\infty}^{+\infty} \exp\left[-\frac{(1+\mathrm{i}C)T^2}{2T_0^2} + \mathrm{i}\omega T\right]\mathrm{d}T \\
&= \sqrt{\frac{1}{1+\mathrm{i}C}} T_0 \exp\left[-\frac{(1-\mathrm{i}C)T_0^2\omega^2}{2(1+C^2)}\right]
\end{aligned} \tag{2-12}
$$

由式(2-12)可得脉冲频谱半峰全宽为

$$
\Delta\nu_{m=1} = \sqrt{\ln 2(1+C^2)} / (\pi T_0) \tag{2-13}
$$

则其时间带宽积为

$$
\Delta T_{m=1} \cdot \Delta\nu_{m=1} = 2\ln 2\sqrt{1+C^2} / \pi \approx 0.441\sqrt{1+C^2} \tag{2-14}
$$

式(2-9)所示啁啾超高斯脉冲的强度自相关函数

$$
\begin{aligned}
G(\tau') &= \int_{-\infty}^{+\infty} \left|U(T)\right|^2 \left|U(T-\tau')\right|^2 \mathrm{d}T \\
&= \int_{-\infty}^{+\infty} \left|\exp\left[-\frac{1+\mathrm{i}C}{2}\left(\frac{T}{T_0}\right)^{2m}\right]\right|^2 \cdot \left|\exp\left[-\frac{1+\mathrm{i}C}{2}\left(\frac{T-\tau'}{T_0}\right)^{2m}\right]\right|^2 \mathrm{d}T
\end{aligned} \tag{2-15}
$$

在 $m=1$ 时(即高斯脉冲)，其强度自相关为

$$
\begin{aligned}
G(\tau')_{m=1} &= \int_{-\infty}^{+\infty} \left|U(T)\right|^2 \left|U(T-\tau')\right|^2 \mathrm{d}T = \int_{-\infty}^{+\infty} \exp\left[-\frac{T^2}{T_0^2} - \frac{(T-\tau')^2}{T_0^2}\right]\mathrm{d}T \\
&= T_0\sqrt{\pi}\exp\left(-\frac{\tau'^2}{2T_0^2}\right)
\end{aligned} \tag{2-16}
$$

其半峰全宽为

$$
\Delta\tau'_{\mathrm{G},m=1} = 2\sqrt{2\ln 2}T_0 = \sqrt{2}\Delta T_{m=1} \tag{2-17}
$$

是相应高斯脉冲半峰全宽的 $\sqrt{2}$ 倍。

式(2-9)所示啁啾超高斯脉冲的自相关脉冲傅里叶变换

$$
\begin{aligned}
\tilde{U}_{\mathrm{s}}(\omega,\tau') &= \int_{-\infty}^{+\infty} U(T)U(T-\tau')\exp(\mathrm{i}\omega T)\mathrm{d}T \\
&= \int_{-\infty}^{+\infty} \exp\left[-\frac{1+\mathrm{i}C}{2}\left(\frac{T}{T_0}\right)^{2m}\right]\exp\left[-\frac{1+\mathrm{i}C}{2}\left(\frac{T-\tau'}{T_0}\right)^{2m}\right]\exp(\mathrm{i}\omega T)\mathrm{d}T
\end{aligned} \tag{2-18}
$$

在 $m=1$ 时(即高斯脉冲)，其自相关脉冲的傅里叶变换

$$
\begin{aligned}
\tilde{U}_{\mathrm{s},m=1}(\omega,\tau') &= \int_{-\infty}^{+\infty} U(T)U(T-\tau')\exp(\mathrm{i}\omega T)\mathrm{d}T \\
&= \sqrt{\frac{\pi}{1+\mathrm{i}C}} T_0 \exp\left[-\frac{\omega^2 T_0^4/(1+\mathrm{i}C) + \mathrm{i}2T_0^2\omega\tau' + (1+\mathrm{i}C)\tau'^2}{4T_0^2}\right]
\end{aligned} \tag{2-19}
$$

式(2-9)所示啁啾超高斯脉冲的自相关频谱函数

$$\begin{aligned} G(\omega) &= \int_{-\infty}^{+\infty} \left| \tilde{U}_{\mathrm{s}}(\omega, \tau') \right|^2 \mathrm{d}\tau' \\ &= \int_{-\infty}^{+\infty} \left| \int_{-\infty}^{+\infty} \exp\left[-\frac{1+\mathrm{i}C}{2} \left(\frac{T}{T_0} \right)^{2m} \right] \exp\left[-\frac{1+\mathrm{i}C}{2} \left(\frac{T-\tau'}{T_0} \right)^{2m} \right] \exp(\mathrm{i}\omega T) \mathrm{d}T \right|^2 \mathrm{d}\tau' \end{aligned} \tag{2-20}$$

在 $m=1$ 时(即高斯脉冲)，脉冲自相关频谱为

$$\begin{aligned} G(\omega)_{m=1} &= \int_{-\infty}^{+\infty} \left| \tilde{U}_{\mathrm{s},m=1}(\omega, \tau') \right|^2 \mathrm{d}\tau' \\ &= \pi T_0^3 \exp\left[-\frac{T_0^2 \omega^2}{2(1+C^2)} \right] \end{aligned} \tag{2-21}$$

其自相关频谱的半峰全宽

$$\Delta \nu_{\mathrm{s},m=1} = \sqrt{2\ln 2(1+C^2)} \,/\, (\pi T_0) = \sqrt{2} \Delta \nu_{m=1} \tag{2-22}$$

是相应高斯脉冲频谱半峰全宽的 $\sqrt{2}$ 倍。则

$$\Delta \lambda_{\mathrm{s},m=1} = \frac{c}{\nu_{\mathrm{s},m=1}^{\ \ 2}} \Delta \nu_{\mathrm{s},m=1} = \frac{c\sqrt{\ln 2(1+C^2)}}{2\sqrt{2}\pi T_0 \nu^2} = \frac{\Delta \lambda_{m=1}}{2\sqrt{2}} \tag{2-23}$$

式中，c 是真空中的光速。由此可以得到自相关脉冲的时间带宽积

$$\Delta \nu_{\mathrm{s},m=1} \cdot \Delta \tau_{\mathrm{G},m=1} = 4\ln 2\sqrt{1+C^2} \,/\, \pi = 2\Delta \nu_{m=1} \cdot \Delta T_{m=1} \tag{2-24}$$

是相应高斯脉冲时间带宽积的 2 倍。

式(2-11)、式(2-15)、式(2-18)和式(2-20)在 $m\neq1$ 时很难解析求解，作者采用数值分析方法得到了 m=2, 3 时超高斯脉冲的自相关特性，并与原脉冲相应参量作了比较。若无特殊说明，数值计算采用时域窗口为(– 40，40)，采样点数为 1024。

2.2.1　啁啾脉冲的自相关特性曲线随 *m* 和 | *C* | 的变化

以超高斯脉冲为例，图 2-3 所示是 $m=2$、$C=0$ 时超高斯脉冲的时域波形及其强度自相关曲线，横坐标是时间和时间延迟(对自相关曲线表示时延)，归一化到脉冲半宽度 T_0，纵坐标是归一化强度；实线所示是超高斯脉冲的时域波形，点线所示是其强度自相关曲线。下文图 2-5、图 2-9 与本图坐标相同。由图 2-3 可见，脉冲时域波形在 $T=0$ 附近非常平坦，波形前后沿下降速度快；脉冲自相关曲线在 $T=0$ 附近比前者稍尖锐，前后沿下降速度比前者缓慢；脉冲自相关曲线半峰全宽比脉冲时域半峰全宽明显要宽。当锐度参量 m 增加时，脉冲强度自相关曲线边缘稍内缩，单边的曲线逐渐近似为直线，半峰全宽稍窄，与脉冲时域波形的平顶展宽，前后沿下降速度加快。当啁啾参量$|C|$增加时，脉冲强度自相关曲线与脉冲时域波形均保持不变。

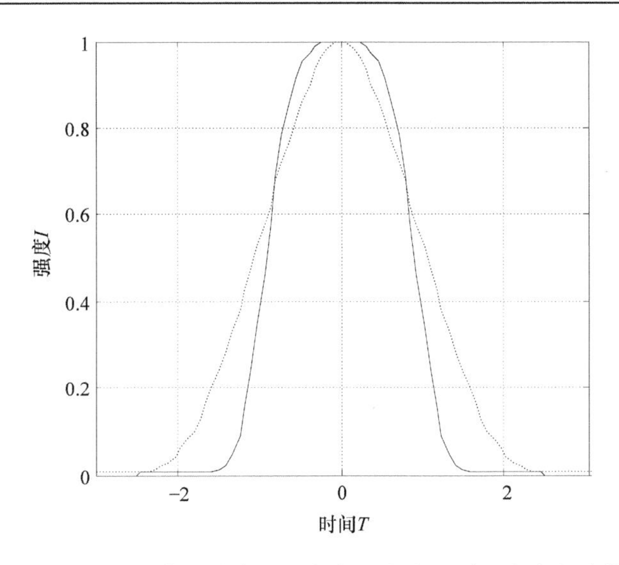

图 2-3　$m=2$、$C=0$ 时超高斯脉冲的时域波形(实线)及其强度自相关曲线(点线)

图 2-4 所示是当 $m=2$、$C=0$ 时超高斯脉冲频谱曲线及其自相关频谱曲线，横坐标是频率，归一化到 $1/T_0$，纵坐标是归一化强度。实线所示是超高斯脉冲频谱曲线，点线所示是其自相关频谱曲线。图 2-6～图 2-8、图 2-10 与本图坐标相同。

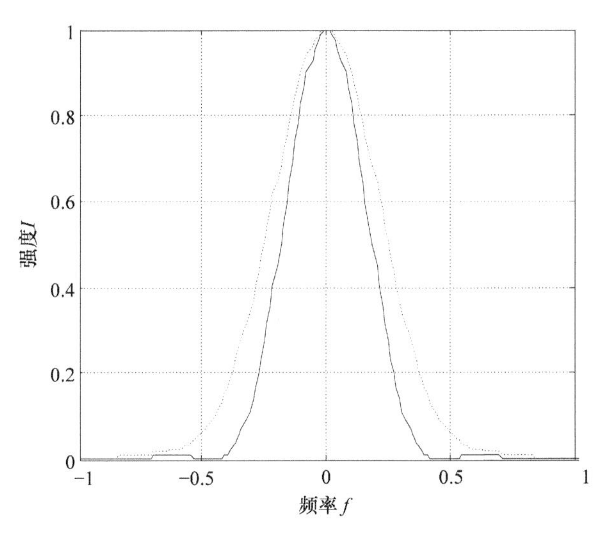

图 2-4　$m=2$、$C=0$ 时超高斯脉冲的频谱曲线(实线)及其自相关频谱曲线(点线)

由图 2-4 可见，脉冲频谱曲线及其自相关频谱曲线在 $f=0$ 附近几乎重合；脉冲频谱曲线前后沿比自相关频谱曲线的下降速度快得多，在脉冲频谱曲线边缘出

现振荡结构，而自相关频谱曲线边缘没有振荡结构；脉冲自相关频谱曲线宽度比脉冲频谱曲线宽度宽。当锐度参量 m 和啁啾参量 $|C|$ 分别增加时，脉冲的频谱曲线继续展宽，边缘的振荡结构随之增大；自相关频谱曲线亦随之展宽，其边缘仍然较平滑。

表 2-1 所示是数值计算得到的超高斯脉冲及其自相关特性随 m 的变化。表 2-1 分别给出了超高斯脉冲特性及其自相关特性的三个参量：时域半峰全宽、频谱半峰全宽和时间带宽积。由表 2-1 可得，m=1 时，数值计算得到的超高斯脉冲及其自相关特性参量与解析结果完全一致；超高斯脉冲的时域半峰全宽、频谱半峰全宽和时间带宽积及其自相关频谱半峰全宽和自相关时间带宽积都随 m 的增大而增加，增加速度逐渐减小；自相关时域半峰全宽随之减小，减小速度逐渐减小。自相关特性参量与脉冲特性参量的比值随之减小且速度变缓。

表 2-1　数值计算得到的超高斯脉冲及其自相关特性

	脉冲特性			自相关特性		
	时域半峰全宽	频谱半峰全宽	时间带宽积	时域半峰全宽	频谱半峰全宽	时间带宽积
m=1，C=0	$1.6651T_0$	$0.2650/T_0$	0.4413	$2.3548T_0$	$0.3748/T_0$	0.8825
m=2，C=0	$1.8249T_0$	$0.3741/T_0$	0.6827	$2.1053T_0$	$0.5053/T_0$	1.0639
m=3，C=0	$1.8815T_0$	$0.4061/T_0$	0.7641	$2.058T_0$	$0.5443/T_0$	1.1201

2.2.2　脉冲噪声对超高斯脉冲及其自相关特性的影响

作者用 SHG-FROG 脉冲分析仪研究脉冲的传输特性时发现，当存在较强的色散波噪声时，脉冲的自相关特性变化比较复杂。为了更深入地研究脉冲自相关特性，本书在原超高斯脉冲时域波形的边缘加入一个幅值较小的超高斯脉冲模拟色散波噪声，并通过数值计算给出了脉冲噪声对脉冲及其自相关特性的影响规律。

包含脉冲噪声的啁啾超高斯脉冲的归一化包络电场可表示为

$$U(T)=\exp\left[-\frac{1+\mathrm{i}C}{2}\left(\frac{T}{T_0}\right)^{2m}\right]+a'\exp\left[-\frac{1+\mathrm{i}C}{2}\left(\frac{T-b'}{T_0}\right)^{2m}\right],\quad -\infty<T<\infty \tag{2-25}$$

式中，a' 为脉冲噪声幅值系数，b' 为超高斯脉冲与脉冲噪声的时间间隔。式(2-25)所示脉冲的频谱及其自相关特性难于解析求解，作者用数值计算得到了其变化规律。

图 2-5 所示是当 m=2、C=0、a'=0.3 和 b'=3 时超高斯脉冲的时域波形及其强度自相关曲线，实线所示是超高斯脉冲的时域波形，点线所示是其强度自相关曲线。图 2-6 所示是当 m=2、C=0、a'=0.3 和 b'=3 时超高斯脉冲频谱曲线及其自相关频谱曲线，实线所示是超高斯脉冲频谱曲线，点线所示是其自相关频谱曲线。

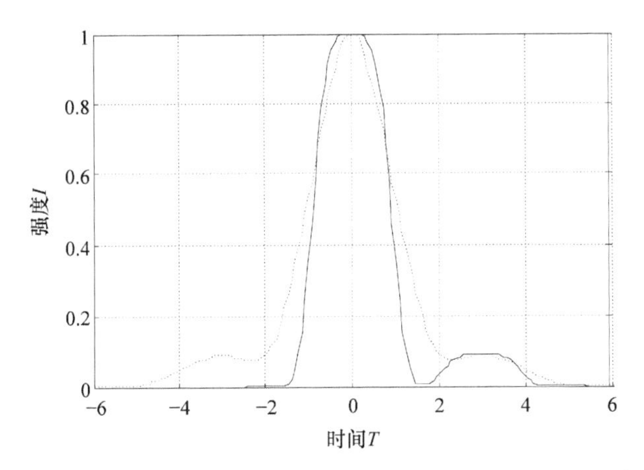

图 2-5　当 m=2、C=0、a' =0.3 和 b' =3 时超高斯脉冲的时域波形(实线)及其强度自相关曲线(点线)

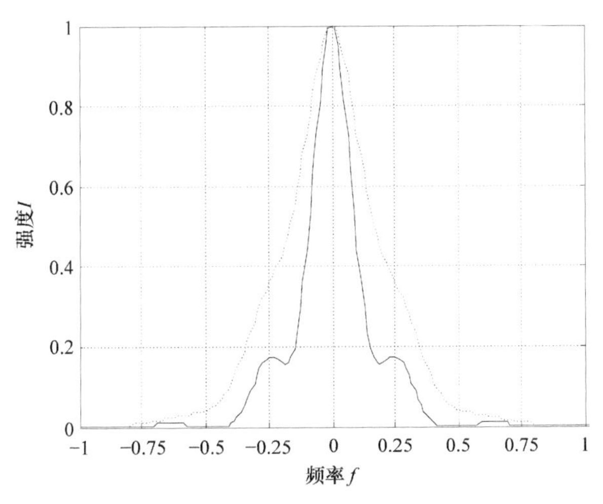

图 2-6　当 m=2、C=0、a' =0.3 和 b' =3 时超高斯脉冲频谱(实线)及其自相关频谱(点线)

由图 2-5 可见，在 T=3 处有幅值 a' =0.3 的脉冲噪声时，原脉冲时域波形几乎不变，时域宽度比无脉冲噪声时稍微增宽；脉冲自相关曲线在$|T|$=3 处出现两个基本对称的振荡结构；脉冲自相关曲线半峰全宽比无脉冲噪声时亦稍宽。在 C 和 b' 不变的情况下，当脉冲噪声幅值 a' 增加时，脉冲的时域宽度和自相关曲线宽度随之稍微增宽，自相关曲线上的振荡结构随之增大。

由图 2-6 可见，在 T=3 处有幅值 a' =0.3 的脉冲噪声时，脉冲频谱曲线及其自相关频谱曲线的边缘都出现了对称的振荡结构和底座，自相关频谱曲线边缘振荡结构和底座的位置比脉冲频谱曲线边缘的要高；脉冲的频谱宽度和自相关频谱宽度比无脉冲噪声时稍窄。在 C 和 b' 不变的情况下，当脉冲噪声幅值 a' 增加时，脉冲的频谱宽度和自相关频谱宽度随之减小，两曲线上的振荡结构随之增大。

在 a' 和 b' 相同的情况下，当啁啾参量 $|C|$ 增加时，脉冲的时域宽度和自相关曲线宽度及其曲线形状几乎不变。脉冲的频谱宽度和自相关频谱宽度随之稍微增加，两曲线上的振荡结构随之增大，如图 2-6 和图 2-7 比较所示。

在脉冲噪声幅值 a' 和啁啾参量 $|C|$ 相同的情况下，当脉冲噪声与原脉冲的间隔增大时，脉冲的时域宽度和自相关曲线宽度随之减小，脉冲噪声对脉冲时域波形和自相关曲线形状的影响越来越小。脉冲的频谱和自相关频谱曲线上的振荡结构随之剧烈增大，甚至出现严重的频谱分裂现象，如图 2-6 和图 2-8 比较所示。这为我们使用 SHG-FROG 脉冲分析仪或者基于自相关技术原理的仪器测量脉冲特性提供了重要参考。为了准确测量脉冲特性，一定要选择合适的时域扫描窗口，尽可能使时域扫描窗内没有脉冲噪声。

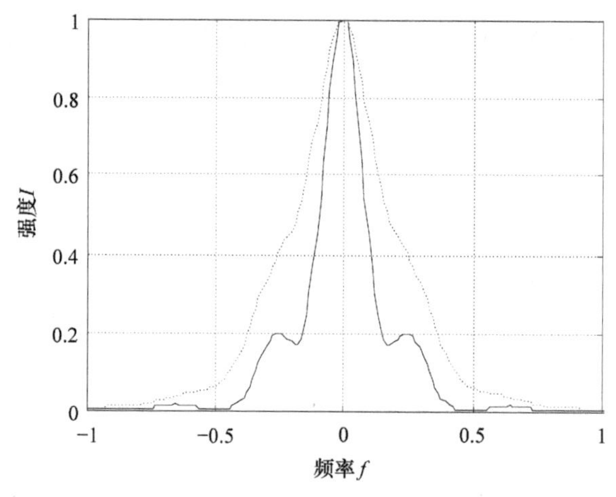

图 2-7　当 m=2、C=|0.5|、a' =0.3 和 b' =3 时超高斯脉冲频谱(实线)及其自相关频谱(点线)

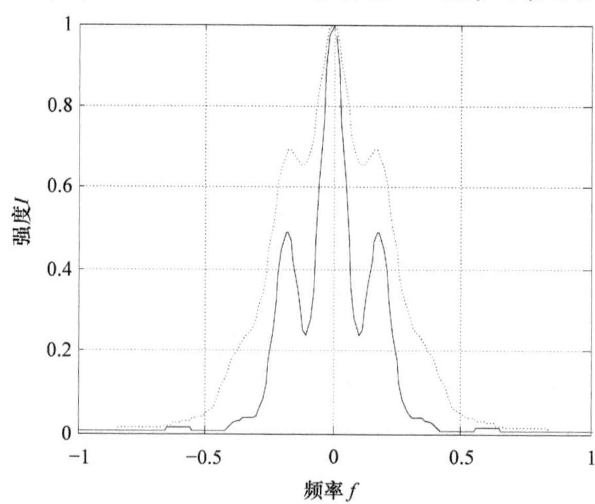

图 2-8　当 m=2、C=0、a' =0.3 和 b' =5 时超高斯脉冲频谱(实线)及其自相关频谱(点线)

2.2.3　仪器随机噪声对超高斯脉冲自相关特性的影响

通常情况下，SHG-FROG 脉冲分析仪存在电荷耦合器件引入的随机噪声。该噪声随器件温度升高而增大，表现为随测量时间的增加而增大。测量时间较长时，该噪声通常为正值，且对测量数据有一定影响。作者采用数值计算方法，以脉冲的二维函数 $I_{\text{test}}(\omega,\tau')=\left|\tilde{E}_{\text{test}}(\omega,\tau')\right|^2$ 数据最大值的 1%～10%(噪声系数)乘以 0～1 的随机函数值模拟随机噪声，分析了随机噪声对脉冲强度自相关和自相关频谱参量的影响。

图 2-9 和图 2-10 分别是在 m=2、C=0 时超高斯脉冲的强度自相关曲线和自相关频谱曲线，实线为无噪声情况，点线为有随机噪声情况，实点为滤除随机噪声后的数据。图 2-9 和图 2-10 数值计算采用时域窗口为(−20, 20)，采样点数为 512，噪声系数为 5%。

由图 2-9 和图 2-10 可以看出，在时域和频谱中心区域，有随机噪声的强度自相关时域宽度和自相关频谱宽度偏离无噪声的情况；在时域和频谱边缘区域，有随机噪声的强度自相关曲线和自相关频谱曲线高于无噪声的情况。当随机噪声幅度增加时，在时域和频谱边缘区域，随机噪声的强度自相关曲线和自相关频谱曲线随之明显增高；在时域和频谱中心区域，随机噪声的强度自相关时域宽度和自相关频谱宽度与无噪声的差别增大；脉冲越宽，采样点数越多，时域窗口越小，时域受影响越大。

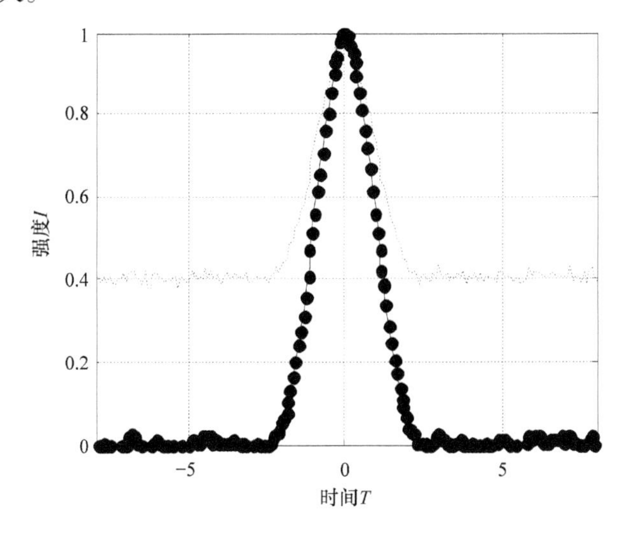

图 2-9　在 m=2、C=0 时超高斯脉冲的强度自相关曲线

实线为无噪声情况，点线为有随机噪声情况，实点为滤除随机噪声后的数据

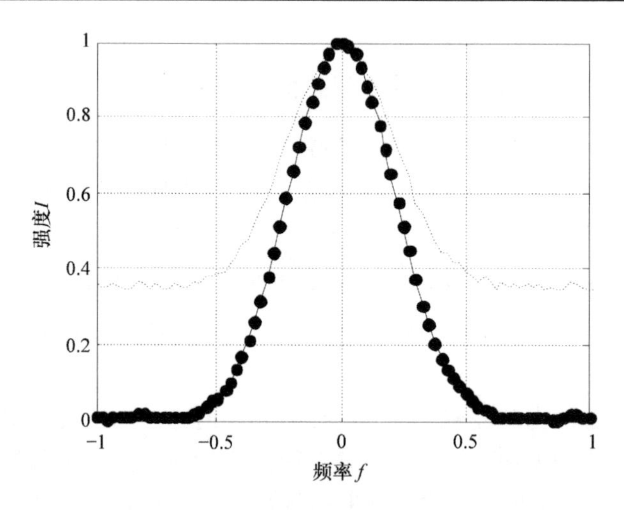

图 2-10 在 m=2、C=0 时超高斯脉冲的自相关频谱曲线

实线为无噪声情况，点线为有随机噪声情况，实点为滤除随机噪声后的数据

考虑到脉冲自相关曲线或自相关频谱曲线边缘(大于 2 倍的半峰全宽时)趋近于零，作者在 FROG 图数据 $I_{\text{test}}(\omega,\tau')$ 边缘各取 n 列或行(本书 n=10)个数据求和，以其平均值作为各点的噪声值，在 FROG 图数据各个数据点滤除噪声值，然后对数据作平滑滤波处理(即第 n–1, n, n+1 三个数据的平均值作为第 n 个数据)且归一到峰值。采取该方法处理后的数据如图 2-9 和图 2-10 中的实点所示，由图可见，滤波处理后的数据与脉冲无噪声的自相关曲线和自相关频谱曲线完全吻合。作者改变时域窗口和采样点数，发现该滤波方法处理后数据依旧与无噪声情况吻合。这启示我们，使用 SHG-FROG 脉冲分析仪或者基于自相关技术原理的仪器测量脉冲特性一定要选择合适的扫描窗口和合适的采样点数，尽可能在测量前就采取必要的消噪处理或者对含有噪声的测量数据做必要的滤波后处理。

2.2.4 滤除随机噪声方法的应用实验

利用SHG-FROG脉冲分析仪对半导体锁模激光器输出的中心波长1548.05nm的脉冲进行了测量分析,得到该脉冲为时域半峰全宽 2.21ps、频谱半峰全宽 2.23nm、时间带宽积 0.615、啁啾参量 $C=-1$ 的高斯脉冲(m=1)。脉冲归一化包络电场可写为

$$U(T)=\exp\left[-\frac{(1+\mathrm{i}C)T^2}{2T_0^2}\right] \tag{2-26}$$

式中，T_0=2.21/1.665ps。图 2-11 和图 2-12 分别是超高斯脉冲(m=1)的强度自相关和自相关频谱曲线，实线为测量得到的实验曲线，点线为利用 2.2.3 节方法去除随

机噪声后的曲线，实点为由式(2-26)理论计算得到的自相关曲线数据。图中纵坐标均为归一化强度，图 2-11 横坐标为时间延迟，单位是皮秒(ps)；图 2-12 横坐标为波长，单位是纳米(nm)，此处波长为倍频波长。由图 2-11 和图 2-12 可见，利用 SHG-FROG 实验测量得到的超高斯脉冲(m=1)的自相关曲线较平滑，时域关于零时延对称，频域关于倍频中心波长对称，表明脉冲时域和频域形状较规则。但是，测量得到的实验曲线包含较大的随机噪声，在实验数据的后处理过程中必须首先滤除噪声。滤除噪声后的曲线与由式(2-26)数值计算得到的数据在时域和频域均吻合得很好，进一步证明了 2.2.3 节滤除随机噪声方法的可靠性和可行性。

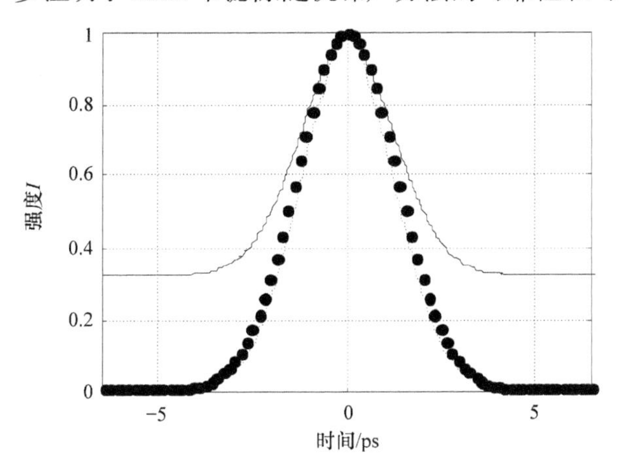

图 2-11　超高斯脉冲的强度自相关曲线

实线为测量得到的实验曲线，点线为去除随机噪声后的曲线，实点为理论计算得到的强度自相关曲线

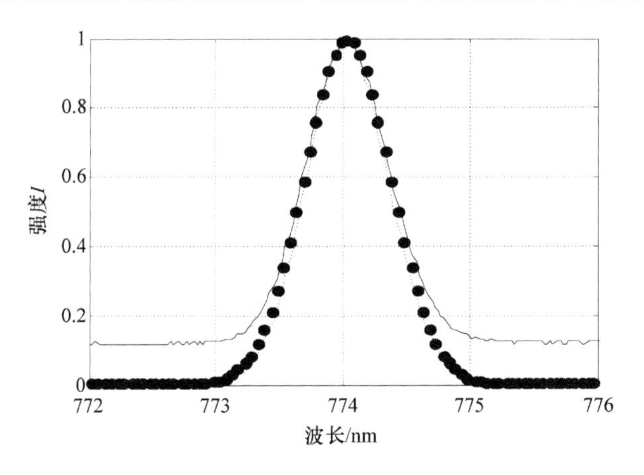

图 2-12　超高斯脉冲的自相关频谱曲线

实线为测量得到的实验曲线，点线为去除随机噪声后的曲线，实点为理论计算得到的自相关频谱曲线

2.3 本章小结

由频率分辨光学门技术原理导出了超高斯脉冲的强度自相关和自相关频谱公式，采用数值分析方法研究了脉冲的自相关特性及其受锐度参量 m、啁啾参量 C、脉冲噪声和随机噪声影响的变化规律，与原脉冲相应参量作了比较，给出了一个有效滤除随机噪声的方法并作了实验验证，为我们使用 SHG-FROG 脉冲分析仪或者基于自相关技术原理的仪器测量脉冲特性提供了重要参考[28-30]。测量脉冲特性一定要选择合适的扫描窗口和合适的采样点数，尽可能在测量前就采取必要的消除随机噪声处理或者对含有噪声的测量数据作必要的滤波后处理，尽可能使时域扫描窗内没有脉冲噪声。

参考文献

[1] 阿戈沃. 非线性光纤光学原理及应用.贾东方, 余震虹, 谈斌, 译. 北京: 电子工业出版社, 2002.

[2] Kane D J, Trebino R. Characterization of arbitrary femtosecond pulses using frequency-resolved optical gating. IEEE J. Quantum Electron., 1993, 29(2): 571～579.

[3] DeLong K W, Trebino R, Hunter J, et al. Frequency-resolved optical gating with the use of second-harmonic generation. J. Opt. Soc. Am. B,1994, 11(11): 2206～2215.

[4] Gallmann L, Steinmeyer G, Sutter D H, et al. Collinear type II second-harmonic-generation frequency-resolved optical gating for the characterization of sub-10-fs optical pulses. Optics Letters, 2000, 25(4): 269～271.

[5] 王兆华, 魏志义, 滕浩, 等. 飞秒激光脉冲的谐波频率分辨光学开关法测量研究. 物理学报, 2003, 52(2): 362～366.

[6] 龙井华, 高继华, 巨养锋, 等. 用 SHG-FROG 方法测量超短光脉冲的振幅和相位. 光子学报, 2002, 31(10): 1292～1296.

[7] DeLong K W, Fittinghoff D N, Trebino R, et al. Pulse retrieval in frequency-resolved optical gating based on the method of generalized projections. Optics Letters, 1994, 19(24): 2152～2154.

[8] Hu J, Zhang G Z, Zhang B G, et al. Using frequency-resolved optical gating to retrieve amplitude and phase of ultrashort laser pulse. Journal of Optoeletronics · Laser, 2002, 13(3): 232～236.

[9] Lacourt P A, Dudley J M, Merolla J M, et al. Milliwatt-peak-power pulse characterization at 1.55 μm by wavelength-conversion frequency-resolved optical gating. Optics Letters, 2002, 27(10): 863～865.

[10] Barry L P, Delburgo S, Thomsen B C, et al. Optimization of optical data transmitters for 40-Gb/s lightwave systems using frequecy resolved optical gating. photon. Tech. Lett., 2002, 14(7): 971～973.

[11] Liu S, Lu D, Zhao L, et al. SHG-FROG characterization of a novel multichannel synchronized

AWG-based mode-locked laser//Conference on Lasers and Electro-Optics, OSA Technical Digest(online)(Optical Society of America, 2017), paper JTh2A.131.

[12] Kane D J. Improved principal components generalized projections algorithm for frequency resolved optical gating//Conference on Lasers and Electro-Optics, OSA Technical Digest (online) (Optical Society of America, 2017), paper STu3I.4.

[13] Hyyti J, Escoto E, Steinmeyer G, et al. Interferometric time-domain ptychography for ultrafast pulse characterization. Opt. Lett., 2017, 42: 2185～2188.

[14] Sidorenko P, Lahav O, Avnat Z, et al. Ptychographic reconstruction algorithm for frequency-resolved optical gating: Super-resolution and supreme robustness. Optica, 2016, 3: 1320～1330.

[15] Heidt A M, Spangenberg D M, Brügmann M, et al. Improved retrieval of complex supercontinuum pulses from XFROG traces using a ptychographic algorithm. Opt. Lett., 2016, 41: 4903～4906.

[16] Ermolov A, Valtna-Lukner H, Travers J, et al. Characterization of few-fs deep-UV dispersive waves by ultra-broadband transient-grating XFROG. Opt. Lett., 2016, 41: 5535～5538.

[17] Fuji T, Shirai H, Nomura Y. Self-referenced frequency-resolved optical gating capable of carrier-envelope phase determination//Conference on Lasers and Electro-Optics, OSA Technical Digest (2016) (Optical Society of America, 2016), paper SM3I.7.

[18] Steinmeyer A. Interferometric FROG for Ultrafast Spectroscopy on the Few-cycle Scale // Conference on Lasers and Electro-Optics, OSA Technical Digest (2016) (Optical Society of America, 2016), paper STu4I.1.

[19] Okamura B, Sakakibara Y, Omoda E, et al. Experimental analysis of coherent supercontinuum generation and ultrashort pulse generation using cross-correlation frequency resolved optical gating (X-FROG). J. Opt. Soc. Am. B, 2015, 32: 400～406.

[20] Itakura R, Kumada T, Nakano M, et al. Frequency-resolved optical gating for characterization of VUV pulses using ultrafast plasma mirror switching. Opt. Express, 2015, 23: 10914～10924.

[21] Snedden E W, Walsh D A, Jamison S P. Revealing carrier-envelope phase through frequency mixing and interference in frequency resolved optical gating. Opt. Express, 2015, 23: 8507～8518.

[22] Hause A, Kraft S, Rohrmann P, et al. Reliable multiple-pulse reconstruction from second-harmonic-generation frequency-resolved optical gating spectrograms. J. Opt. Soc. Am. B, 2015, 32: 868～877.

[23] Li X J, Liao J L, Nie Y M, et al. Unambiguous demonstration of soliton evolution in slow-light silicon photonic crystal waveguides with SFG-XFROG. Opt. Express, 2015, 23: 10282～10292.

[24] 罗时荣, 吕百达, 张彬. 平顶高斯光束与超高斯光束传输特性的比较研究. 物理学报, 1999, 48(8): 1446～1451.

[25] 卿与三, 吕百达. 平顶高斯光束和超高斯光束传输特性的相似性. 强激光与粒子束, 2001, 13(6): 675～678.

[26] 吴建伟, 夏光琼, 吴正茂.超高斯光脉冲在单模光纤中的传输特性. 激光技术, 2003, 27(4): 342～348.

[27] 曹涧秋, 陆启生. 单模光纤中高阶色散对超高斯光脉冲传播的影响. 激光技术, 2006, 30(2): 209～220.

[28] 郑宏军, 刘山亮, 黎昕, 等. 超高斯光脉冲自相关特性研究. 中国激光, 2007, 34(7): 908～914.

[29] Zheng H J, Liu S L, Li X, et al. Autocorrelation characteristics of the double-side exponential pulse with linear chirp. Proc. of SPIE, 2007, 6783: 67834A-1-6.

[30] Zheng H J, Liu S L, Li X. Effects of chirp and noises on autocorrelation characteristics of hyperbolic secant pulse. Proc. of SPIE, 2008, 7136: 71363B-1-5.

第 3 章　啁啾脉冲的线性传输研究

随着激光技术、光纤技术、光放大技术和波分复用等技术的发展，短脉冲的线性传输研究得到了快速发展。然而，由于受到实验条件等各种因素的限制，在以往的脉冲传输与测量实验中[1-33]只是注重输入、输出脉冲的宽度及其变化，没有或者很少研究波形、啁啾和时间带宽积的变化。频率分辨光学门(FROG)测量技术[1, 34-55]可以有效抑制背景，具有较高的动态范围，能够准确测量脉冲的脉宽、谱宽、波形、相位等特征参量信息，可以广泛应用于各种脉冲的测量。就作者所知，目前国内外了解和应用这种测量技术研究脉冲传输的科研人员还为数不多。最近，作者完成了 10GHz 短脉冲在色散平坦光纤中的线性传输实验，利用二次谐波频率分辨光学门(SHG-FROG)脉冲分析仪测量得到了传输前后的短脉冲的 FROG 图、自相关曲线、自相关频谱曲线、波形和相位曲线，以及脉宽、谱宽、啁啾等特征参量信息，将实验测量的这些参量及其传输前后的变化与脉冲的线性传输理论预期进行了比较。结果表明，实验测量结果、数值计算结果与理论预期一致，SHG-FROG 测量技术是探测、分析短脉冲的一种有效工具。在此基础上，进一步研究了线性和非线性啁啾双曲正割脉冲和啁啾双边指数脉冲的线性传输特性，填补了人们对其线性传输规律认识的国际空白，给出了两脉冲频谱宽度和时间带宽积随啁啾变化的表达式，提供了判断脉冲时域波形的一种有效方法。

3.1　啁啾脉冲的线性传输实验与理论分析

3.1.1　脉冲传输前的实验测量

利用 SHG-FROG 脉冲分析仪对半导体锁模脉冲激光器(mode-locked pulse laser)输出的短脉冲进行了测量分析，实验装置如图 3-1 所示。实验所用激光器为德国 U2T 公司的可调谐半导体锁模脉冲激光器 TMLL1550，可调谐波长范围为 1470～1570nm，可调谐重复频率范围为 9.8～10.8GHz，平均输出光功率与输出脉冲的中心波长、偏置电流有关。光隔离器(optical isolator)为武汉光讯公司生产，隔离度大于 61dB，插入损耗为 0.6dB。EDFA 为法国 KPS 公司的掺铒光纤放大器，噪声指数<6dB，增益平坦度<1.5dB，小信号增益>40dB，光放大波段为 1535～1565nm。FROG 分析仪为新西兰南方光子有限公司的 SHG-FROG 脉冲分析仪 HR200，采用铌酸锂($LiNbO_3$)晶体作为非线性介质，时间分辨率即最小时间延迟为 26.66fs。HR200 采用 1200g/mm 的高分辨率反射光栅作为光谱仪，CCD 相邻像素的间隔为 0.023nm，频谱分辨小于 0.05nm。HR200 脉冲分析仪对偏振非常敏感，要求输入功率较高，一般不小于 20mW。为了测量脉冲传输前的特性，图 3-1 所示实验装置中先不接入 12.7km 色散平坦光纤(DFF)。在本书所述实验中，半导体锁模激光器输出的短脉冲中心波长为 1548nm、重复频率为 10GHz、平均输出功率为 0.3mW，经光隔离器后直接进入掺铒光纤放大器，放大到 50mW 后输入 HR200 进行测量和分析。

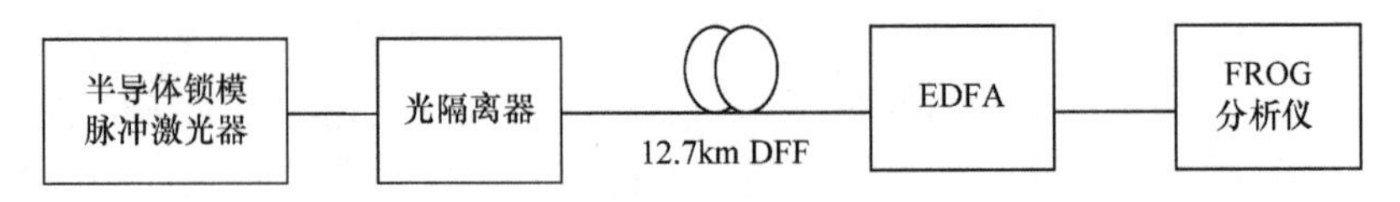

图 3-1　脉冲线性传输实验装置

图 3-2 所示是测量得到脉冲传输前的 FROG 图。图中横坐标表示时间延迟，单位是飞秒(fs)，纵坐标表示自相关脉冲波长，单位是纳米(nm)。由图 3-2 可见，图中可以表现出频谱强度的变化，强度从频谱中心向外依次递减。自相关脉冲频谱强度分布关于零时延和中心波长基本上是对称的，说明测量脉冲的波形和频谱比较规则。

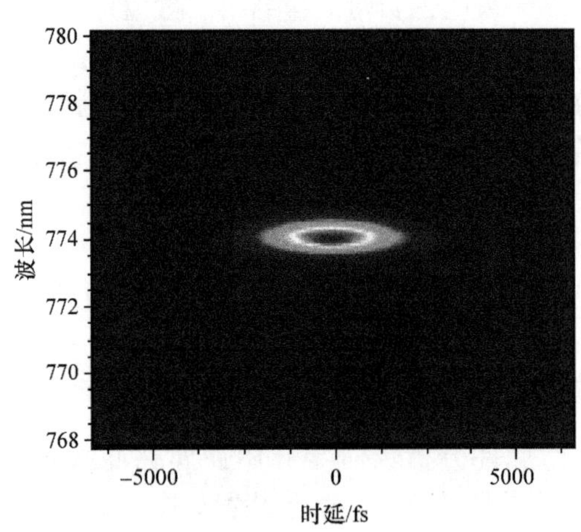

图 3-2　脉冲传输前的 FROG 图

图 3-3 所示是测量脉冲的强度自相关曲线，图中横坐标表示时间延迟，单位是飞秒(fs)，纵坐标表示自相关脉冲强度，任意单位。由图 3-3 可见，测量脉冲的自相关曲线比较平滑且对称、无底座，其半峰全宽为 2.84ps，是由式(2-17)得到的相应高斯脉冲半峰全宽等于 2.84ps 除以 1.414，约为 2.01ps。

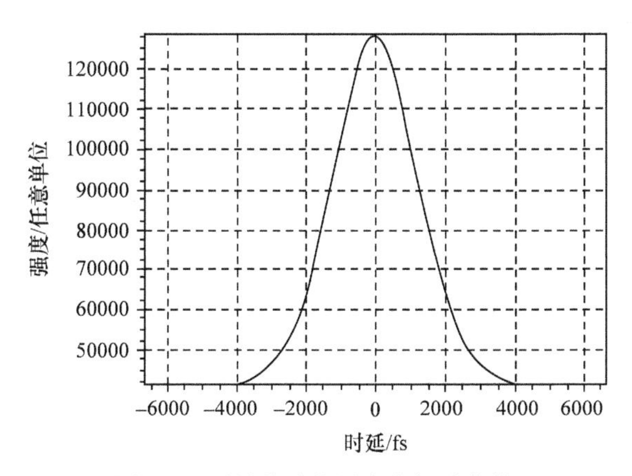

图 3-3　测量脉冲的强度自相关曲线

图 3-4 所示是测量脉冲的自相关频谱曲线，图中横坐标表示自相关脉冲的波长(频率)，单位是纳米(nm)，纵坐标表示自相关频谱强度，任意单位。由图 3-4 可见，测量脉冲的自相关频谱曲线比较平滑且对称、无底座，其半峰全宽为 0.72nm，由式(2-23)得到的相应高斯脉冲频谱半峰全宽是 0.72nm 的 2.828 倍，约为 2.04nm。

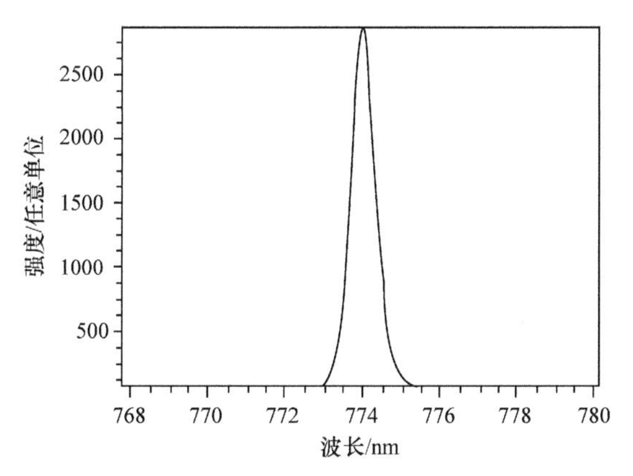

图 3-4　测量脉冲的自相关频谱曲线

图 3-5 所示是实验测量得到的脉冲时域波形和相位曲线，图中横坐标表示时间，单位是皮秒(ps)，左纵坐标表示脉冲归一化强度，右纵坐标表示脉冲相位，单位是弧度(rad)。由图 3-5 可见，测量脉冲有较小的啁啾。根据图 3-5 中的相位-时间曲线和啁啾的定义可以给出实验脉冲的啁啾参量 $C=-1$，时域半峰全宽 2.21ps、频谱半峰全宽 2.23nm 和时间带宽积 0.615 由 HR200 脉冲分析仪(SHG-FROG)直接给出。表 3-1 列出了所测脉冲的有关特性参量。

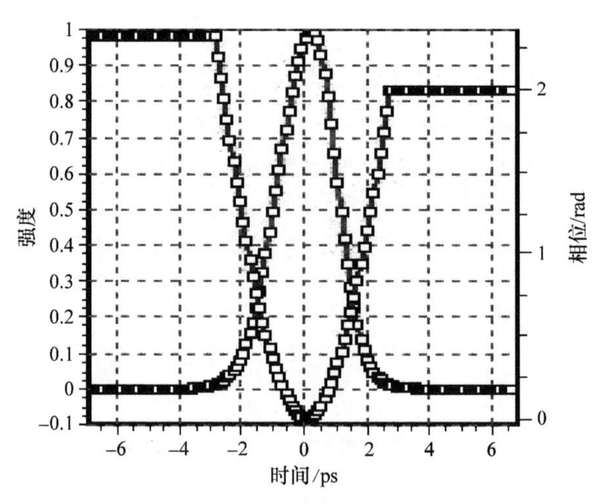

图 3-5 实验测量得到的脉冲时域波形和相位曲线

表 3-1 脉冲传输前的特性参量

方法	时域半峰全宽/ps	频谱半峰全宽/nm	时间带宽积
实验值(SHG-FROG)	2.21	2.23	0.615
实验值(高斯自相关)	2.01	2.04	0.513

由表 3-1 可见，由图 3-3 和图 3-4 脉冲强度自相关曲线和自相关频谱曲线计算得到的高斯脉冲的时域宽度、谱宽度和时间带宽积都比 SHG-FROG 测量得到的实验脉冲的相应值略小。SHG-FROG 测量得到的实验脉冲的脉宽与由强度自相关曲线给出的高斯脉冲的基本一致，然而与由强度自相关曲线计算得到的双曲正割脉冲的半峰全宽 1.83ps 相差较大。由此初步推断，所测脉冲的波形是高斯型的。

3.1.2 实验测量分析方法

针对脉冲的 SHG-FROG 测量新技术，给出并采用了相应的分析新方法。首先，将实验测量得到的脉冲波形、相位等数据从 SHG-FROG 脉冲分析仪导入作者设计的 Matlab 程序中，然后将实验波形数据与常用的脉冲波形(如高斯和双曲正割波形等)进行曲线拟合比较，找到与实验数据最接近的波形；实验得到的脉冲通常带有较大的频率啁啾，啁啾 $\delta\omega=-\dfrac{\partial\phi}{\partial t}$ 通常根据测量得到的脉冲相位曲线即 ϕ-t 曲线给出，这样可以把实验相位数据与线性啁啾相位或其他形式的相位进行曲线拟合比较，找到与实验数据最接近的相位或啁啾形式，确定啁啾参量 C；根据解析或数值求解可以得到的不同波形脉冲的时间带宽积和啁啾参量 C 的表达式，将已经确定的啁啾参量 C 代入该表达式计算得到相应的时间带宽积，再与 SHG-FROG 脉冲分析仪测量得到的脉冲时间带宽积比较，然后找出与实验数据最接近的时间带宽积及其相应波形；最后可将实验数据与数值计算得到的双曲正割和高斯脉冲传输特性作比较，找出实验数据最接近的脉冲特性及其相应波形。综合考虑上述信息，就可以准确地确定实验测量得到的脉冲波形、啁啾等参量。

3.1.3　脉冲传输前的实验测量分析

将实验测量得到的脉冲波形、相位等数据导入 Matlab 程序中进行曲线拟合比较。图 3-6 所示是半导体锁模激光器输出的脉冲传输前实验测量得到的时域波形(a)和相位曲线(b)。图中横坐标是时间，单位皮秒(ps)，图(a)纵坐标是脉冲的归一化强度，图(b)纵坐标是脉冲相位，单位弧度(rad)。图 3-6(a)中实线所示是实验测量得到的脉冲时域波形曲线，点线所示是与实验脉冲波形具有相同时域半峰全宽的高斯脉冲

$$U(0,\tau)=\exp[-0.5(1+\mathrm{i}C)\tau^2] \tag{3-1}$$

的波形曲线，式(3-1)中，时间 $\tau=T/T_0$, T_0=2.21/1.665ps 是脉冲强度 1/e 处的半宽度；脉冲时域半峰全宽为

$$\Delta T=2\sqrt{\ln 2}T_0 \tag{3-2}$$

脉冲的相位是 $\phi=-CT^2/(2T_0^2)$, 脉冲的频率啁啾为 $\delta\omega=-\partial\phi/\partial T=CT/T_0^2$ ，可见，瞬时频率随时间线性变化，称为线性啁啾；C 为线性啁啾参量，若 $C<0$，称为负啁啾或下啁啾；若 $C>0$，称为正啁啾或上啁啾。由相位公式可得，脉冲中心 T=0 处的相位为 0，啁啾参量 C 在数值上等于 $T=T_0$ 处相位的 2 倍，可以由脉冲的相位-时间曲线给出 $C=-1$。点划线所示是线性啁啾双曲正割脉冲

$$U(0,\tau)=\mathrm{sech}(\tau)\exp(-0.5\mathrm{i}C\tau^2) \tag{3-3}$$

的波形曲线，式(3-3)中，时间 $\tau=T/T_0$ 是归一化时间，T_0 = 2.21/1.763ps。由图 3-6(a)可见，输入脉冲时域波形与高斯曲线非常吻合，仅脉冲后沿下降较高斯曲线下降稍快；脉冲前后沿远远偏离了双曲正割波形。

图 3-6(b)实线所示是实验测量得到的输入脉冲的相位曲线，点线和点划线所示分别是线性啁啾参量 $C=-1$ 时高斯脉冲式(3-1)和双曲正割脉冲式(3-3)对应的相位曲线。由图 3-6(b)可见，实验测量得到的相位曲线与高斯脉冲的相位曲线非常吻合，偏离了双曲正割脉冲的相位曲线；脉冲相位在中心附近关于 $T=0$ 基本

对称，脉冲前沿的相位比后沿的稍大，与脉冲时域波形曲线的后沿下降快相联系。本书采用相位曲线拟合方法得到了线性啁啾参量 $C=-1$ 与采用当 $T=T_0$ 时线性啁啾 C 在数值上是相位的两倍的方法($\phi=-CT^2/(2T_0^2)$)得到的结果一致。

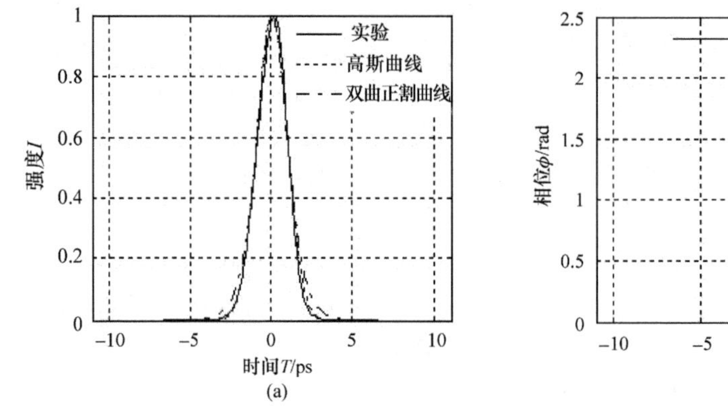

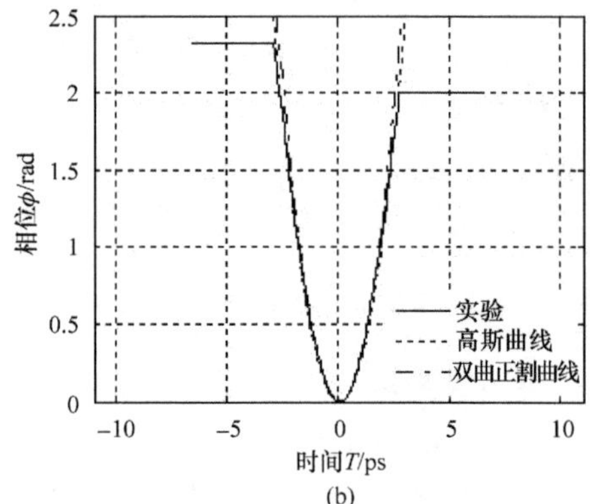

图 3-6　脉冲传输前实验测量得到的时域波形(a)和相位曲线(b)

将 $C=-1$ 代入线性啁啾高斯脉冲的时间带宽积随频率啁啾参量 C 变化的公式计算得到 $\Delta T\cdot\Delta\nu=0.441\sqrt{1+C^2}=0.624$，与实验测量得到的 0.615 基本一致。双曲正割脉冲时间带宽积随频率啁啾参量 C 变化公式不能解析得到，作者用数值计算方法得到了其时间带宽积公式，将 $C=-1$ 代入得到 $\Delta T\cdot\Delta\nu=0.461|C|+0.315=0.776$，与实验测量得到的 0.615 差别很大。

由此可见，可以进一步判定半导体锁模激光器输出的脉冲在传输前是具有负线性啁啾 $C=-1$ 的高斯短脉冲，而不是双曲正割脉冲。

3.1.4　脉冲线性传输后的实验测量分析

在图 3-1 所示实验装置中的光隔离器与掺铒光纤放大器之间接入 12.7km 的色散平坦光纤。半导体锁模激光器输出的 1548nm 短脉冲经 12.7km 色散平坦光纤传输后输入掺铒光纤放大，将放大后脉冲输入 SHG-FROG 脉冲分析仪进行测量和分析。实验所用色散平坦光纤在 1548nm 波长处的群速色散系数 $\beta_2=-0.22\text{ps}^2/\text{km}$。图 3-7 所示是传输 12.7km 后的自相关曲线，图中横坐标表示时间延迟，单位是飞秒(fs)，纵坐标表示自相关脉冲强度，任意单位。由测量得到的自相关曲线图 3-7 和 SHG-FROG 测量得到的实验脉冲时域波形和相位曲线图 3-8(图中坐标与图 3-5 相同)可见，脉冲经过 12.7km 色散平坦光纤传输后明显展宽。自相关曲线的半峰全宽由输入脉冲的 2.84ps 展宽到 8.95ps，SHG-FROG 测量得到的实验脉冲的时域半峰全宽由输入脉冲的 2.21ps 增加到 6.88ps，与由自相关曲线得到的相应高斯脉冲宽度 6.33ps 吻合。输出脉冲的频谱宽度与输入脉冲的大致相同。输出脉冲的时间带宽积 1.85 是输入脉冲 0.615 的 3 倍，远大于高斯脉冲变换极限值 0.441，是脉冲展宽的结果。从实验脉冲的相位曲线图 3-8 可见，经 12.7km 色散平坦光纤传输后脉冲啁啾明显增加。由图 3-8 所示的相位曲线和相位公式 $\phi=-CT^2/(2T_0^2)$ 给出的脉冲啁啾参量 $C=-4$，是输入脉冲的 4 倍。由不同方法得到的输出脉冲特性参量如表 3-2 所示。

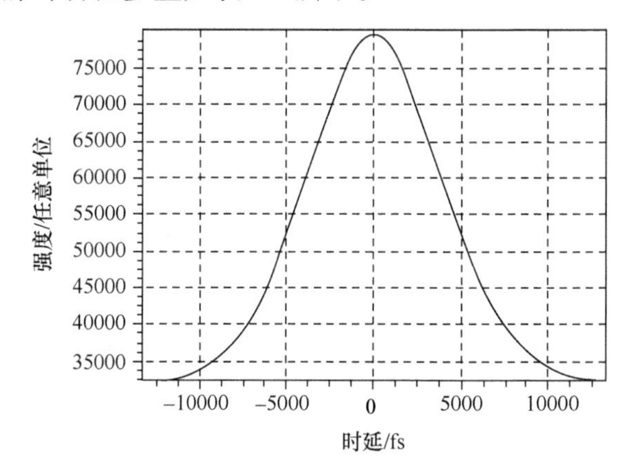

图 3-7　脉冲传输 12.7km 后的强度自相关曲线

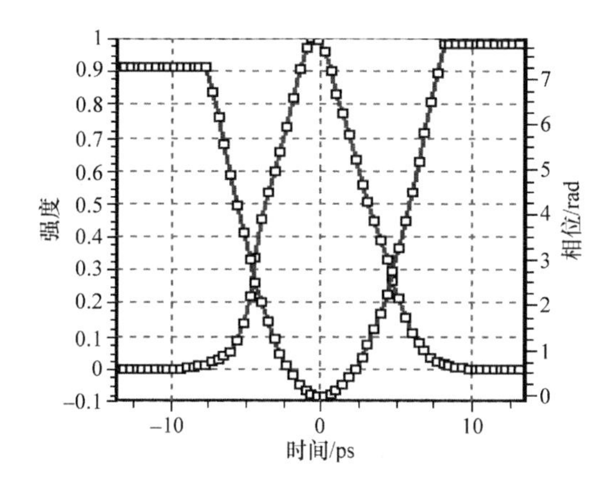

图 3-8　脉冲传输 12.7km 后的波形和相位曲线

将 SHG-FROG 脉冲分析仪测量得到的脉冲经过 12.7km 色散平坦光纤传输后的时域波形和相位曲线等数据导入 Matlab 程序进行曲线拟合比较。图 3-9 所示是半导体锁模激光器输出的脉冲经过 12.7km 色散平坦光纤传输后实验测量得到的时域波形(a)和相位曲线(b)，本图与图 3-6 相应坐标相同。

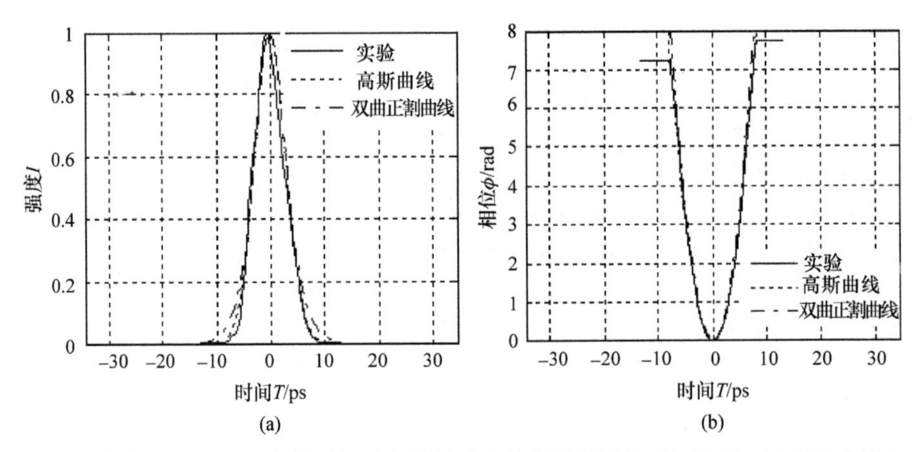

图 3-9　脉冲经过 12.7km 色散平坦光纤传输后实验测量得到的时域波形(a)和相位曲线(b)

由图 3-9(a)可见，输入脉冲时域波形与高斯曲线基本吻合，仅脉冲前沿下降比高斯曲线下降稍快；脉冲前后沿远远偏离了双曲正割波形。图 3-9(b)实线所示是实验测量得到的输入脉冲的相位曲线，点线和点划线所示分别是线性啁啾参量 $C=-4.2$ 时高斯脉冲式(3-1)和双曲正割脉冲式(3-3)对应的相位曲线。由图 3-9(b)可见，实验测量得到的相位曲线与高斯脉冲的相位曲线非常吻合，偏离了双曲正割脉冲的相位曲线；脉冲相位在中心附近关于 $T=0$ 基本对称，脉冲前沿的相位比后沿的稍小，这与脉冲时域波形曲线的前沿下降快有关。

本节采用相位曲线拟合方法得到的线性啁啾参量 $C=-4.2$ 与采用当 $T=T_0$ 时线性啁啾 C 在数值上是相位的两倍的方法得到的结果 $C=-4$ 基本一致。曲线拟合方法因考虑到整个相位曲线的变化情况，结果更准确，优于相位公式 $\phi=-CT^2/(2T_0^2)$ 的方法。

将 $C=-4.2$ 代入线性啁啾高斯脉冲的时间带宽积公式，计算得到 $\Delta T\cdot\Delta\nu=0.441\sqrt{1+C^2}=1.904$，与实验测量得到的 1.85 基本一致；将 $C=-4.2$ 代入双曲正割脉冲时间带宽积公式得到 $\Delta T\cdot\Delta\nu=0.461|C|+0.315=2.251$，与实验值 1.85 差别很大。

由此可见，可以初步判定半导体锁模激光器输出的脉冲经过 12.7km 色散平坦光纤传输后是具有更大负线性啁啾 $C=-4.2$ 的高斯脉冲，而不是双曲正割脉冲。

3.1.5　高斯脉冲线性传输理论

在光纤非线性、高阶色散和光纤损耗的影响可以忽略的情况下，脉冲包络电场 $U(z,T)$ 在光纤中传输满足的方程(1-21)修正为

$$\mathrm{i}\frac{\partial U}{\partial z}=\frac{\beta_2}{2}\frac{\partial^2 U}{\partial T^2}\tag{3-4}$$

式中，$\beta_2 = \mathrm{d}^2\beta / \mathrm{d}\omega^2$ 是光纤的群速色散系数。$U(z,T)$ 的傅里叶变换 $\tilde{U}(z,\omega)$ 满足

$$\mathrm{i}\frac{\partial \tilde{U}}{\partial z} = -\frac{\beta_2}{2}\omega^2\tilde{U} \tag{3-5}$$

其解为

$$\tilde{U}(z,\omega) = \tilde{U}(0,\omega)\exp\left(\frac{\mathrm{i}}{2}\beta_2\omega^2 z\right) \tag{3-6}$$

$$\tilde{U}(0,\omega) = \int_{-\infty}^{+\infty}\exp\left[-\frac{(1+\mathrm{i}C)T^2}{2T_0^2} + \mathrm{i}\omega T\right]\mathrm{d}T = \sqrt{\frac{2\pi}{1+\mathrm{i}C}}T_0\exp\left[-\frac{(1-\mathrm{i}C)T_0^2\omega^2}{2(1+C^2)}\right] \tag{3-7}$$

是式(3-1)所示线性啁啾高斯脉冲的傅里叶变换。由式(3-7)可得，其频谱的半峰全宽$\left(\left|\tilde{U}(0,\omega)\right|^2 = \frac{1}{2}\left|\tilde{U}(0,\omega)\right|_{\max}^2 \text{处的频谱宽度}\right)$为

$$\Delta\nu = \sqrt{\ln 2(1+C^2)}/(\pi T_0) \quad \text{或} \quad \Delta\lambda = c\sqrt{\ln 2(1+C^2)}/(\pi T_0\nu^2) \tag{3-8}$$

由式(3-2)和式(3-8)可得式(3-1)所示线性啁啾高斯脉冲的时间带宽积为

$$\Delta T\cdot\Delta\nu = 0.441\sqrt{1+C^2} \tag{3-9}$$

将式(3-7)代入式(3-6)，得

$$\tilde{U}(z,\omega) = \sqrt{\frac{2\pi}{1+\mathrm{i}C}}T_0\exp\left[-\frac{(1-\mathrm{i}C_z)T_0^2\omega^2}{2(1+C^2)}\right] \tag{3-10}$$

式中

$$C_z = C + (1+C^2)\beta_2 z / T_0^2 \tag{3-11}$$

由式(3-10)可得，脉冲经过传输后的谱宽

$$\Delta\nu_z = \sqrt{\ln 2(1+C^2)}/(\pi T_0) = \Delta\nu \tag{3-12}$$

与式(3-8)所示的传输前谱宽完全一样，即脉冲的谱宽 $\Delta\nu$ 在传输过程中保持不变，相位和啁啾参量随着传输距离的变化而变化。式(3-10)可以写为

$$\tilde{U}(z,\omega) = \sqrt{\frac{2\pi}{1+\mathrm{i}C}}T_0\exp\left[-\frac{T_z^2\omega^2}{2(1+\mathrm{i}C_z)}\right] \tag{3-13}$$

式中

$$T_z^2 = T_0^2[(1+C\beta_2 z/T_0^2)^2 + \beta_2^2 z^2/T_0^4] \tag{3-14}$$

在 z 处的脉冲包络电场写为

$$U(z,T)=\frac{1}{2\pi}\int_{-\infty}^{+\infty}\tilde{U}(z,\omega)\exp(-\mathrm{i}\omega T)\mathrm{d}\omega=\sqrt{\frac{1+\mathrm{i}C_z}{1+\mathrm{i}C}}\frac{T_0}{T_z}\exp\left[-\frac{(1+\mathrm{i}C_z)T^2}{2T_z^2}\right] \tag{3-15}$$

由此可见，虽然在线性传输过程中群速色散(β_2)不会影响脉冲的频谱，但是会改变脉冲的相位、啁啾和宽度。

3.1.6　脉冲线性传输的数值分析与讨论

为了更准确地判断实验测量得到的脉冲波形，本书将与传输前脉冲具有相同时域半峰全宽 2.21ps 和线性啁啾参量 $C=-1$ 的高斯脉冲式(3-1)和双曲正割脉冲式(3-3)分别作为输入脉冲，求解脉冲线性传输方程(3-4)的归一化方程

$$\mathrm{i}\frac{\partial u}{\partial \xi}+\frac{1}{2}\frac{\partial^2 u}{\partial \tau^2}=0 \tag{3-16}$$

式中，u 为脉冲归一化包络电场，ξ 是归一化传输距离，$\tau=\dfrac{T}{T_0}$ 为归一化时间。

双曲正割脉冲式(3-3)作为输入脉冲时，不能解析求解脉冲线性传输方程(3-16)。本书采用分步傅里叶方法，根据式(3-16)对其在色散平坦光纤中的传输 12.7km 的情况进行了数值计算。输入双曲正割脉冲的线性啁啾 $C=-1$，时域半峰全宽 2.21ps，频谱半峰全宽 2.749nm，时间带宽积 0.761，$L_D=\dfrac{T_0^2}{|\beta_2|}=\dfrac{(2.21/1.763)^2}{0.22}=7.143\text{km}$，12.7km 色散平坦光纤约为 1.78 个色散长度。数值计算得到脉冲传输后时域半峰全宽为 8.05ps，频谱半峰全宽为 2.749nm，时间带宽积为 2.772，如表 3-2(数值计算，双曲正割)所示。双曲正割脉冲的数值结果与实验数据差别较大。

同时，作者把高斯脉冲式(3-1)作为输入脉冲进行了数值计算，输入高斯脉冲的线性啁啾 $C=-1$，时域半峰全宽 2.21ps，频谱半峰全宽 2.25nm，时间带宽积 0.624，$L_D=\dfrac{T_0^2}{|\beta_2|}=\dfrac{(2.21/1.665)^2}{0.22}=8.008\text{km}$，12.7km 色散平坦光纤约为 1.59 个色散长度。数值计算得到脉冲传输后时域半峰全宽为 6.72ps，频谱半峰全宽为 2.25nm，时间带宽积为 1.897，如表 3-2(数值计算，高斯)所示，高斯脉冲的数值结果与实验数据基本一致，与表 3-2(理论值，高斯)所示的理论值基本吻合。表 3-2 中输出脉冲的理论值是根据传输理论式(3-11)等和实验有关参量计算得到的。将实验所用色散平坦光纤的长度 12.7km、群速色散系数 $\beta_2=-0.22\text{ps}^2/\text{km}$ 和输入脉冲啁啾参量 $C=-1$、半宽度 $T_0=\Delta T/1.665=2.21/1.665=1.33(\text{ps})$代入式(3-11)、式(3-12)和式(3-14)，计算得到色散平坦光纤输出脉冲的啁啾参量 $C_z=-4.16$，与实验值−4.2 基本吻合，谱宽 $\Delta\nu=2.23$ nm 保持不变，脉宽 $\Delta T_z=6.72$ ps。

表 3-2　脉冲经过 12.7km 色散平坦光纤线性传输后的特性参量

方法	时域半峰全宽/ps	谱域半峰全宽/nm	时间带宽积
实验值(SHG-FROG)	6.88	2.14	1.85
实验值(高斯自相关)	6.33	2.21	1.751
理论值(高斯)	6.72	2.23	1.875
数值计算(高斯)	6.72	2.25	1.897
数值计算(双曲正割)	8.05	2.749	2.772

由表 3-2 得到，SHG-FROG 测量得到的传输后脉冲特性参量与高斯脉冲线性传输理论值和数值计算结果以及自相关测量值基本一致，与双曲正割脉冲线性传输数值计算值差别较大。由此可见，半导体锁模激光器输出的具有较大负线性啁啾的高斯短脉冲，经过 12.7km 色散平坦光纤传输后，成为脉宽更宽、负线性啁啾更大的高斯脉冲，而不是双曲正割脉冲。虽然高斯脉冲的谱宽在传输过程中保持不变，但是其时域宽度、相位和啁啾参量随着传输距离的变化而变化。

3.2　啁啾双曲正割脉冲的线性传输

自从在非线性传输中提出光孤子以来[56]，关于双曲正割脉冲的研究受到了广泛关注，但是其研究工作大多限于非线性传输中[1, 9-15,56-86]。双曲正割脉冲在光纤中线性传输特性的变化规律很难解析给出，人们对其线性传输规律的认识受到很大限制。鉴于此，作者数值研究了啁啾双曲正割脉冲的线性传输特性。

3.2.1　线性啁啾双曲正割脉冲的特性参量

线性啁啾双曲正割脉冲的归一化包络电场可表示为

$$u(0,\tau)=\mathrm{sech}(\tau)\exp(-0.5\mathrm{i}C\tau^{2}) \tag{3-17}$$

式中，$\tau=T/T_0$ 为归一化时间。由式(3-17)可得脉冲时域半峰全宽(FWHM)为 $\Delta\tau=2\ln(1+\sqrt{2})\approx1.7627$，比归一化线性啁啾高斯脉冲

$$u(0,\tau)=\exp(-0.5(1+\mathrm{i}C)\tau^{2}) \tag{3-18}$$

的时域半峰全宽 $2\sqrt{\ln 2}\approx1.6651$ 稍大。

式(3-17)所示线性啁啾双曲正割脉冲的频率啁啾为

$$\delta\omega = -\frac{\partial \phi}{\partial \tau} = C\tau \tag{3-19}$$

式中，$\phi = -0.5C\tau^2$ 为脉冲相位。由式(3-19)可见，瞬时频率从脉冲前沿到后沿随时间线性变化，称为线性频率啁啾。C 是线性频率啁啾参量，当 $C>0$ 时，称为上啁啾或正啁啾；当 $C<0$ 时，称为下啁啾或负啁啾。

式(3-17)所示线性啁啾双曲正割脉冲的频谱为

$$u(0,\omega) = \int_{-\infty}^{\infty} u(0,\tau)\exp(\mathrm{i}\omega\tau)\mathrm{d}\tau = \int_{-\infty}^{\infty} \mathrm{sech}(\tau)\exp(-0.5\mathrm{i}C\tau^2)\exp(\mathrm{i}\omega\tau)\mathrm{d}\tau \tag{3-20}$$

当 $C=0$ 时，对式(3-20)积分可得无啁啾双曲正割脉冲的频谱

$$u(0,\omega) = \sqrt{\frac{\pi}{2}}\,\mathrm{sech}\left(\frac{\pi\omega}{2}\right) \tag{3-21}$$

由式(3-21)可得无啁啾双曲正割脉冲的频谱半峰全宽

$$\Delta\nu_0 = \frac{\Delta\omega_0}{2\pi} = \frac{2\ln(1+\sqrt{2})}{\pi^2} \approx 0.1786$$

约为无啁啾高斯脉冲频谱半峰全宽 $\sqrt{\ln 2}/\pi \approx 0.265$ 的 2/3。则无啁啾双曲正割脉冲的时间带宽积为

$$\Delta\tau \cdot \Delta\nu_0 = \frac{4[\ln(1+\sqrt{2})]^2}{\pi^2} \approx 0.3148$$

比无啁啾高斯脉冲时间带宽积 $2\ln 2/\pi \approx 0.4413$ 略小。

当 $C\neq 0$ 时，至今无人能够解析求解式(3-20)，也就无法给出其频谱及其半峰全宽、时间带宽积与线性啁啾参量 C 的关系。作者通过数值计算和线性拟合给出了双曲正割脉冲时间带宽积与线性啁啾参量 C 之间的变化规律，如图 3-10 所示，横坐标是线性啁啾参量 C，纵坐标是频谱半峰全宽，归一化到脉冲的 $1/T_0$。

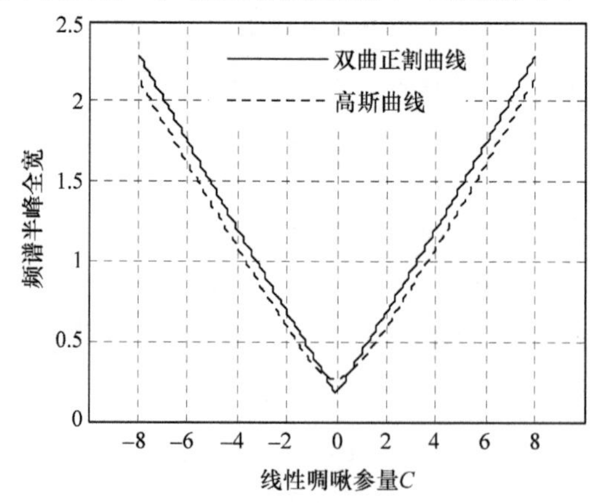

图 3-10　脉冲频谱半峰全宽随线性啁啾参量 C 的变化

图中实线所示是数值计算得到的双曲正割脉冲谱宽随啁啾 C 的变化曲线，与

$$\Delta\nu = 0.2615|C| + 0.1786,\quad |C| \leqslant 8 \tag{3-22a}$$

的曲线吻合，最大误差为 0.0207。点线所示是数值计算得到的啁啾高斯脉冲谱宽随啁啾参量 C 的变化曲线，与高斯脉冲谱宽 $\Delta\nu = \sqrt{\ln 2\cdot(1+C^2)}/\pi = 0.265\sqrt{1+C^2}$ 非常吻合，最大误差为 10^{-4}。双曲正割脉冲频谱半峰全宽随啁啾参量$|C|$的增大而线性增加，比高斯脉冲的变化快；对于相同的$|C|$值，当$|C|\leqslant 0.47$时，双曲正割脉冲频谱半峰全宽小于高斯脉冲的；当$|C|>0.47$ 时，双曲正割脉冲频谱半峰全宽大于高斯脉冲的，$|C|$越大，两者差距越大。则双曲正割脉冲时间带宽积随啁啾 C 变化的表达式

$$\Delta\tau\cdot\Delta\nu = 0.461|C| + 0.3148,\quad |C| \leqslant 8 \tag{3-22b}$$

与高斯脉冲的时间带宽积 $2\ln 2\sqrt{1+C^2}/\pi$ 比较可得，双曲正割脉冲时间带宽积随啁啾参量$|C|$的增大而线性增加，比高斯脉冲的变化快；对于相同的$|C|$值，当$|C|\leqslant 0.38$ 时，双曲正割脉冲时间带宽积小于高斯脉冲的；当$|C|>0.38$ 时，双曲正割脉冲时间带宽积大于高斯脉冲的。式(3-22)给出了判断脉冲时域波形的一种有效方法。

3.2.2　双曲正割脉冲线性传输的数学模型

脉冲在单模光纤中线性传输满足归一化传输方程(3-16)，$\mathrm{i}\dfrac{\partial u}{\partial\xi} + \dfrac{1}{2}\dfrac{\partial^2 u}{\partial\tau^2} = 0$。作者采用分步傅里叶方法按照上述方程对双曲正割脉冲的线性传输特性进行了数值研究，并与高斯脉冲线性传输特性进行了比较。输入双曲正割脉冲和高斯脉冲分别为式(3-17)和式(3-18)。数值计算时，时域窗口设置为(−40，40)，采样点数为 256，采样时间间隔为 80/256=0.3125。若无特殊说明，本节计算都采用上述设置。

3.2.3　双曲正割脉冲线性传输的特性

3.2.3.1　线性啁啾双曲正割脉冲时域波形的演化

图 3-11 所示是不同线性啁啾和传输距离时双曲正割脉冲的时域波形。实线所示是数值计算得到的双曲正割脉冲线性传输波形，虚线所示是与其具有相同时域 FWHM 的双边指数脉冲波形，点线所示是相应的高斯脉冲波形，点划线所示是相应的双曲正割脉冲波形。横坐标是归一化时间，纵坐标是时域波形的归一化强度。本节中图 3-12、图 3-13 和图 3-15 与本图坐标相同。

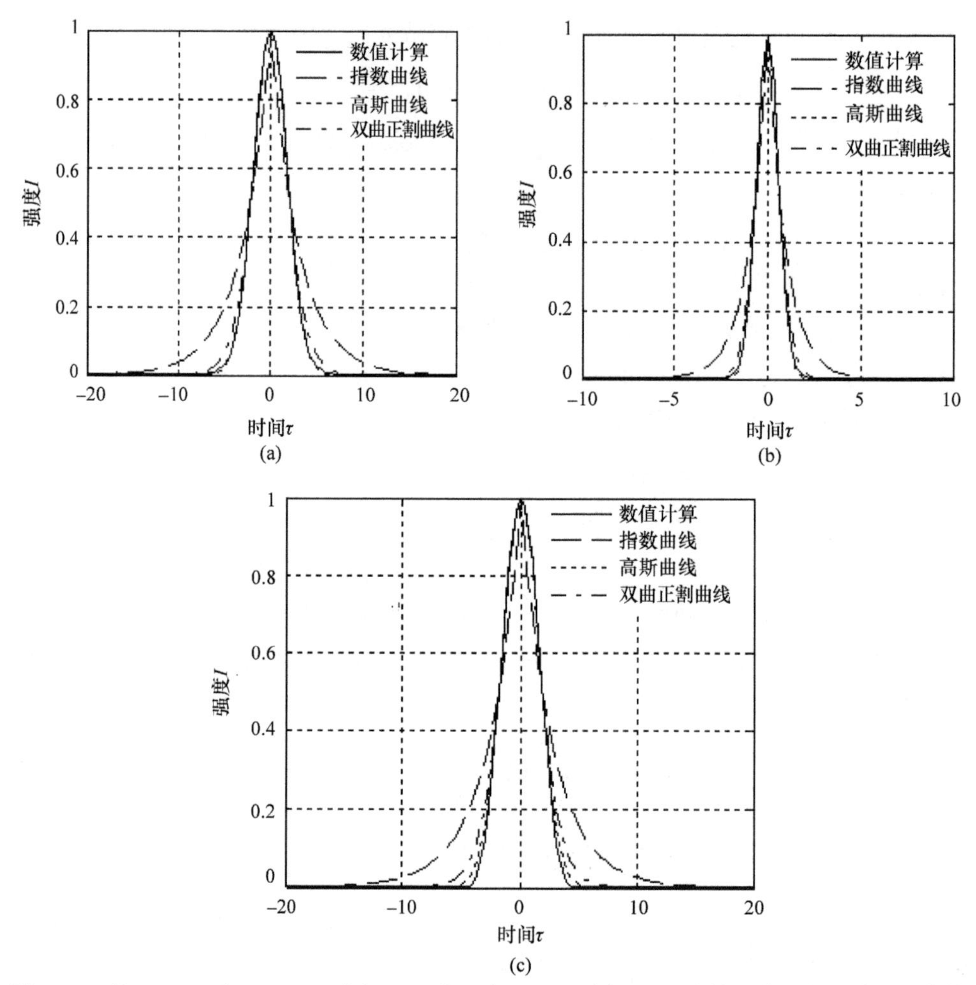

图 3-11　当 $C=-1$(a)和 $C=1.2$(b)时在 $\xi=1$ 处、当 $C=1$(c)时在 $\xi=2$ 处的双曲正割脉冲线性传输的时域波形

数值结果表明，在 $C<-0.1$ 或 $C\geqslant 1.1$ 情况下，线性啁啾双曲正割脉冲在 $\xi=1$ 处演化为近高斯脉冲，在随后的传输中保持不变。啁啾参量 $|C|$ 值越大，双曲正割脉冲演化过程中波形边缘的色散波越大。图 3-11(a)和(b)实线所示分别是双曲正割脉冲 $C=-1$ 和 $C=1.2$ 时传输到 $\xi=1$ 的时域波形。在 $0.7\leqslant C\leqslant 1.1$ 情况下，啁啾脉冲在 $\xi=1$ 处演化为近双曲正割脉冲，在 $\xi=2$ 处演化为近高斯脉冲，在随后的传输中保持不变。图 3-11(c)实线所示是双曲正割脉冲 $C=1$ 时传输到 $\xi=2$ 的时域波形。

在 $0.4<C<0.7$ 情况下，啁啾脉冲在 $\xi=1$ 处演化为近高斯脉冲，在 $\xi=2$ 处演化为近双曲正割脉冲，$\xi=3$ 处演化为近高斯脉冲，在随后的传输中保持不变。

图 3-12(a)、(b)和(c)实线所示是双曲正割脉冲在 $C=0.5$ 时分别传输到 $\xi=1$、2 和 3 处的时域波形。

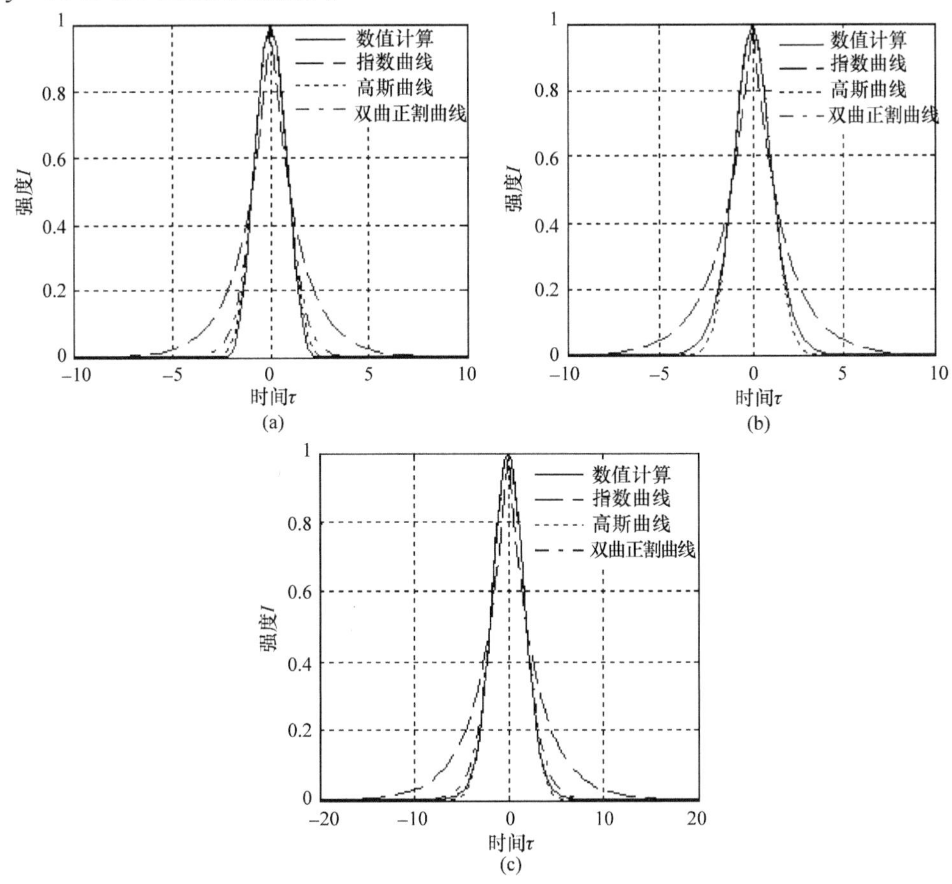

图 3-12　当 $C=0.5$ 时，双曲正割脉冲线性传输到 $\xi=1$ (a)、$\xi=2$ (b)和 $\xi=3$ (c)处的时域波形

在 $0.3 \leqslant C \leqslant 0.4$ 情况下，啁啾脉冲在 $\xi=1$ 处演化为近高斯脉冲，在 $\xi=3$ 处演化为近双曲正割脉冲，$\xi=6$ 处演化为近高斯脉冲，在随后的传输中保持不变。在 $0.1<C<0.3$ 情况下，啁啾脉冲在 $\xi=1$ 处演化为近高斯脉冲，在 $\xi=4$ 处演化为近双曲正割脉冲，$\xi=15$ 处演化为近高斯脉冲，在随后的传输中保持不变。

在 $0 \leqslant |C| \leqslant 0.1$ 情况下，啁啾脉冲最后将演化为近双曲正割脉冲；$|C|$越小，脉冲时域波形越趋近双曲正割曲线。在 $C=0$ 情况下，脉冲在 $\xi=1$ 处演化为近高斯脉冲，在 $\xi=10$ 处演化为近双曲正割脉冲，$\xi=15$ 处的脉冲时域波形与双曲正割曲线完全重合，在随后的传输中保持不变。图 3-13(a)和(b)实线所示是双曲正割脉冲在 $C=0$ 时分别传输到 $\xi=15$ 和 100 处的时域波形，为了精确计算，在时域窗口(−320，320)，采样点数为 2048。

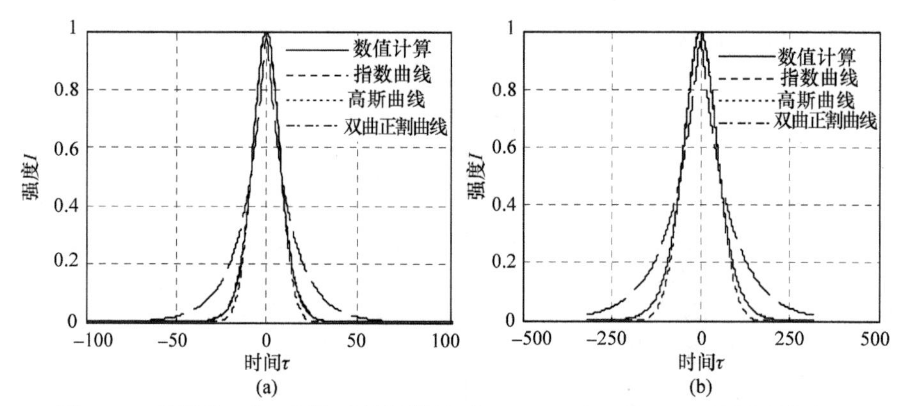

图 3-13　无啁啾双曲正割脉冲线性传输到 $\xi=15$ (a)和 $\xi=100$ (b)处的时域波形

上述结果表明，无啁啾双曲正割光脉冲是一种稳定的线性传输模式；在弱线性啁啾 $(0<|C|\leqslant 0.1)$ 作用下，其时域波形在线性传输时演化成近双曲正割型；当 $|C|>0.1$ 时，双曲正割光脉冲演化成近高斯型。

3.2.3.2　线性啁啾脉冲时域宽度随传输距离的变化

图 3-14 所示是脉冲时域半峰全宽(temporal FWHM)随归一化传输距离的变化。实线 1～3 分别对应双曲正割脉冲线性啁啾 $C=0$、1 和−1 的情况，点线 4～6 分别对应高斯脉冲线性啁啾 $C=0$、1 和−1 的情况。横坐标是归一化传输距离，纵坐标是脉冲时域半峰全宽，归一化到脉冲半宽度 T_0。由图 3-14 可见，$C\leqslant 0$ 时，双曲正割脉冲时域宽度随传输距离的增加而单调展宽，随啁啾参量 C 的增加而减小；在 $C>0$ 时，脉宽首先经历一个初始压缩阶段，然后随传输距离和啁啾参量 C 的增加而展宽；啁啾参量 C 越大，在 $\xi=1$ 附近的初始压缩越强烈。这表明初始正啁啾能诱导双曲正割脉冲压缩。

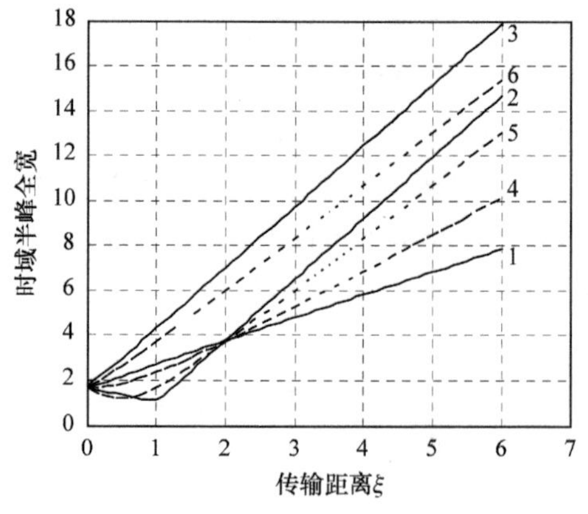

图 3-14　脉冲时域半峰全宽随归一化传输距离的变化

实线 1～3 分别对应双曲正割脉冲 $C=0$、1 和−1 的情况，点线 4～6 分别对应高斯脉冲 $C=0$、1 和−1 的情况

当$|C|\geqslant 0.5$时，双曲正割脉冲时域宽度随传输距离和啁啾参量 C 展宽速度比高斯脉冲的要大；当$|C|<0.5$ 时，比高斯脉冲的要小。负啁啾对应的脉冲展宽速度比正啁啾的快。这表明双曲正割脉冲在$|C|\geqslant 0.5$时受啁啾的影响比高斯脉冲更敏感，负啁啾对脉冲展宽的影响比正啁啾大。因为脉冲压缩对应峰值的增加，所以两脉冲峰值$|u|_{max}$随传输距离和啁啾参量 C 的变化与时域宽度的变化相反。由于线性传输时脉冲频谱不变，所以两脉冲时间带宽积随传输距离和啁啾参量 C 的变化与其脉宽的变化类似。

3.2.3.3　非线性啁啾对脉冲时域波形的影响

作者数值研究了非线性啁啾双曲正割脉冲在线性传输中的时域波形变化，当脉冲边缘加入非线性啁啾微扰时，首次观察到了双曲正割脉冲在线性传输中的波形分裂现象。非线性啁啾对应的非线性相位是

$$\tau\leqslant -2.1875,\phi=a;\quad \tau\geqslant 2.1875,\phi=b;\quad |\tau|<2.1875,\ \phi=-C\tau^2/2 \qquad (3\text{-}23)$$

式中，a 和 b 都是常数；数值计算中，时域窗口为 80，采样点数为 256，采样时间间隔为 0.3125，$|\tau|=7\times0.3125=2.1875$ 对应脉冲的边缘。这种非线性相位是从高斯脉冲线性传输实验中导出实验波形和相位数据，然后进行数据拟合得到。进一步研究发现，在脉冲波形近似为零时，实验仪器就认为相位保持前一个数据不变。也就是说，在式(3-23)中，$|\tau|\geqslant 2.1875$对应的相位分别保持 a 和 b 不变，这并非实验的实际情况，而是$|\tau|\geqslant 2.1875$对应的脉冲波形已经近似为零，仪器就把相位处理为保持 a 和 b 不变而已。虽然这种啁啾并非实验的实际情况，然而，若双曲正割脉冲具有这种啁啾，其线性传输时会导致波形分裂。作者认为这种啁啾导致波形分裂的现象是比较有意义的，故对其进行了讨论。

图 3-15 所示是当 a=3、b=2.5 和 $C=-1$ 时非线性啁啾双曲正割脉冲线性传输到$\xi=6$处的时域波形。数值结果表明，当 $a>b$ 时，双曲正割脉冲在线性传输中出现了不对称的时域波形分裂现象，其中前峰比后峰低，如图 3-15 所示；当 $a<b$ 时，出现另一种不对称的时域波形分裂现象，其中前峰比后峰高；当 $a=b$ 时，出现对称的时域波形分裂现象，其前峰和后峰一样高。啁啾参量$|C|$值越大，波形分裂和波形边缘的色散波越强。

图 3-16 所示是 a=3、b=2.5 和 $C=-1$ 时非线性啁啾双曲正割脉冲的频谱。图中横坐标是频率，归一到 $1/T_0$；纵坐标是频谱强度，任意单位。由图 3-16 可得，非线性啁啾改变了双曲正割脉冲的初始频谱，使初始频谱分裂为双峰结构；因为不同频率分量在光纤中传输速度不同，所以两个谱峰之间的时间间隔随着传播距离的增加而增大，最终导致脉冲在时域上的分裂现象。啁啾参量$|C|$越大，

脉冲频谱变化越剧烈，导致时域波形的分裂越严重。当$-1.7<|C|<2.3$时，在非线性啁啾作用下，高斯脉冲时域波形几乎不变，这是因为在$|\tau|=2.1875$附近，高斯入射脉冲时域波形趋近零的速度比双曲正割脉冲的更快，高斯脉冲频谱受非线性啁啾的影响比双曲正割脉冲的小。由于脉冲频谱受非线性啁啾的影响随$|C|$的增加而增大，高斯脉冲时域波形在$C\leqslant -1.7$或$C\geqslant -2.3$时也出现了波形分裂。

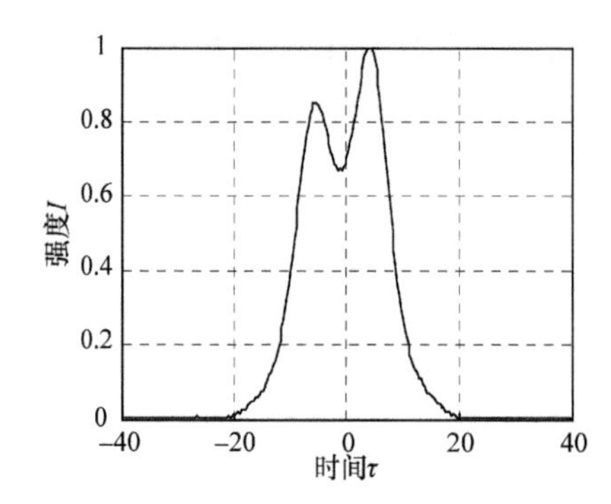

图 3-15　当a=3、b=2.5 和$C=-1$时非线性啁啾双曲正割脉冲传输到$\xi=6$处的时域波形

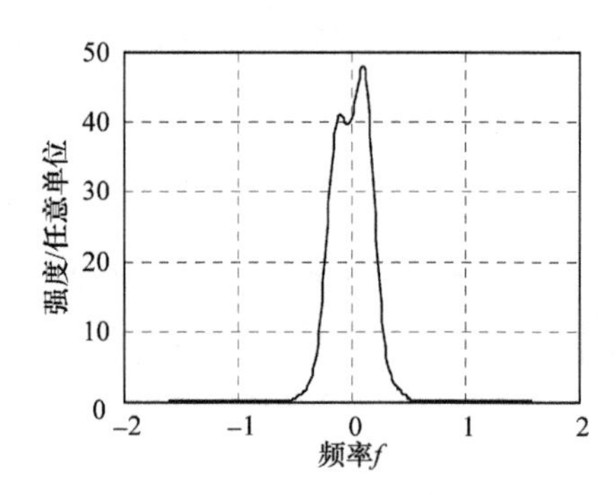

图 3-16　a=3、b=2.5 和$C=-1$时非线性啁啾双曲正割脉冲的频谱

3.3　啁啾指数脉冲的线性传输

以往对指数脉冲的研究仅限于$T\geqslant 0$的情况[87-89]，双边指数脉冲($-\infty<T<\infty$)的传输特性难于解析研究。鉴于此，作者数值研究了啁啾双边指数脉冲的线性传输特性。

3.3.1　啁啾指数脉冲的特性

线性啁啾指数脉冲的归一化电场为

$$U(0,T)=\exp\left(-\frac{|T|}{T_0}\right)\exp\left(-\frac{\mathrm{i}CT^2}{2T_0^2}\right),\qquad -\infty<T<\infty \tag{3-24}$$

其中，T为时间，T_0是e^{-2}强度处的半宽度，C为线性啁啾参量。则脉冲时域半峰全宽是

$$\Delta T=\ln 2T_0\approx 0.693T_0 \tag{3-25}$$

其相位是$\phi=-\dfrac{CT^2}{2T_0^2}$，频率啁啾为$\delta\omega=-\dfrac{\partial\phi}{\partial T}=\dfrac{CT}{T_0^2}$，$C$为线性啁啾参量，$C>0$时称

为上啁啾或者正啁啾，C<0 时称为下啁啾或者负啁啾。

线性啁啾指数脉冲的频谱为

$$\begin{aligned}U(0,\omega)&=\int_{-\infty}^{\infty}U(0,T)\exp(\mathrm{i}\omega T)\mathrm{d}T\\&=\int_{-\infty}^{\infty}\exp\left(-\frac{|T|}{T_0}\right)\exp\left(-\frac{\mathrm{i}CT^2}{2T_0^2}\right)\exp(\mathrm{i}\omega T)\mathrm{d}T\end{aligned}\tag{3-26}$$

C=0 时，积分方程(3-26)得到无啁啾指数脉冲的频谱

$$U(0,\omega)=\frac{2T_0}{1+\omega^2T_0^2}\tag{3-27}$$

由方程(3-27)得到无啁啾指数脉冲的频谱半峰全宽为

$$\Delta\nu_0=\frac{\sqrt{\sqrt{2}-1}}{\pi T_0}\approx 0.205/T_0\tag{3-28}$$

由方程(3-25)和(3-28)得到无啁啾指数脉冲的时间带宽积为

$$\Delta T\cdot\Delta\nu_0=\ln 2T_0\cdot\frac{\sqrt{\sqrt{2}-1}}{\pi T_0}\approx 0.142\tag{3-29}$$

比无啁啾高斯脉冲时间带宽积 0.441 和无啁啾双曲正割脉冲的 0.315 小。

$C\neq 0$ 时，方程(3-26)难于解析求解，通过数值求解方程(3-26)得到了脉冲频谱和频谱半峰全宽随啁啾变化的表达式。图 3-17 所示为数值计算得到的指数脉冲和高斯脉冲的频谱半峰全宽随啁啾参量 C 的变化。横坐标是啁啾参量 C；纵坐标是频谱半峰全宽，归一化到 $1/T_0$。实线所示为指数脉冲频谱半峰全宽随啁啾 C 的变化曲线，与

$$\Delta\nu=(0.001|C|^3-0.0203|C|^2+0.2779|C|+0.205)/T_0,\quad |C|\leqslant 8\tag{3-30}$$

吻合得很好。点线所示是高斯脉冲的相应曲线，与 $\Delta\nu=\sqrt{\ln 2(1+C^2)}/\pi t_0$ 完全一致。由图 3-17 得到，指数脉冲的频谱半峰全宽在 $|C|\leqslant 0.3$ 时小于高斯脉冲的；在 $0.3<|C|\leqslant 3.52$ 时大于高斯脉冲的；在 $|C|>3.52$ 时小于高斯脉冲的，啁啾 $|C|$ 越大，两脉冲频谱半峰全宽的差别越大。

从方程(3-25)和(3-30)得到指数脉冲时间带宽积随啁啾 C 的变化关系

$$\Delta T\cdot\Delta\nu=0.0007|C|^3-0.0141|C|^2+0.1926|C|+0.142,\quad |C|\leqslant 8\tag{3-31}$$

与高斯脉冲的时间带宽积 $2\ln 2\sqrt{1+C^2}/\pi$ 比较得到，两脉冲时间带宽积随啁啾 $|C|$ 的增加而增大，指数脉冲增加得慢。

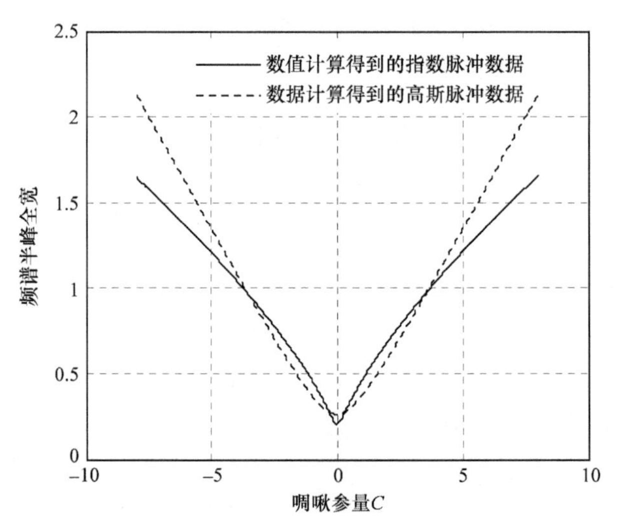

图 3-17 脉冲谱宽随线性啁啾 C 的变化

3.3.2 初始线性啁啾指数脉冲的线性传输特性

采用分步傅里叶方法对啁啾指数脉冲的线性传输特性进行数值研究，并与啁啾高斯脉冲的相应特性作比较。脉冲线性传输满足归一化方程(3-16)，$\mathrm{i}\frac{\partial u}{\partial \xi}+\frac{1}{2}\frac{\partial^2 u}{\partial \tau^2}=0$，归一化输入指数和高斯脉冲分别为

$$u(0,\tau)=\exp(-|\tau|)\exp(-0.5\mathrm{i}C\tau^2) \tag{3-32}$$

和

$$u(0,\tau)=\exp[-0.5(1+\mathrm{i}C)\tau^2] \tag{3-33}$$

式中，$\tau=\frac{T}{T_0}$ 为归一化时间。

3.3.2.1 线性啁啾脉冲时域宽度和时间带宽积随距离的变化

图 3-18 所示是时域半峰全宽随归一化传输距离的变化。实线 1～3 分别是 C=0、1 和−1 时指数脉冲的相应曲线，点线 4～6 分别是 C=0、1 和−1 时高斯脉冲的相应曲线。本图与图 3-14 坐标相同。数值结果发现，C<0.5 时，指数脉冲的时域半峰全宽随距离增加而单调增大，随啁啾 C 的增加而减小；$C\geqslant 0.5$ 时，指数脉冲的时域半峰全宽先经历一个初始减小阶段，后随距离和啁啾 C 的增加而增大；啁啾 C 值越大，指数脉冲的时域半峰全宽在 $\xi=1$ 附近的初始减小的变化越大，这表明初始正啁啾能诱导指数脉冲压缩。

当 $C<0$ 和 $C\geqslant 0.5$ 时，指数脉冲时域半峰全宽随传输距离和啁啾的变化比高斯脉冲的变化快；$0\leqslant C<0.5$ 时，比高斯脉冲的变化慢。负啁啾脉冲比正啁啾脉冲展宽得快，这表明负啁啾对两脉冲时域宽度的影响比正啁啾要大。

由于两脉冲的频谱在线性传输过程中保持不变，两线性啁啾脉冲的时间带宽积随传输距离的变化与时域宽度的变化类似。

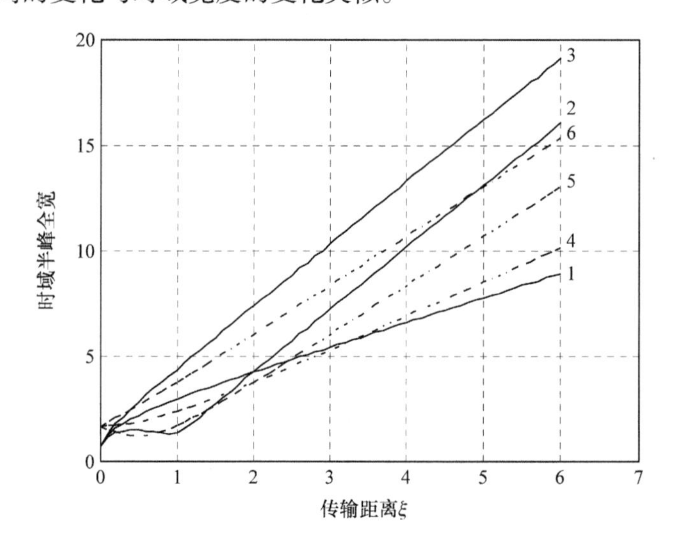

图 3-18　脉冲时域半峰全宽随归一化传输距离的变化

实线 1～3 所示对应指数脉冲 $C=0$、1 和–1 的情况，点线 4～6 所示对应高斯脉冲 $C=0$、1 和–1 的情况

3.3.2.2　线性啁啾脉冲时域波形峰值随距离的变化

图 3-19 所示是两线性啁啾脉冲峰值随归一化传输距离的变化。横坐标是归一化传输距离，纵坐标是脉冲归一化时域波形峰值。实线 1～3 所示分别对应指数脉冲 $C=0$、1 和–1 的情况，点线 4～6 所示分别对应高斯脉冲 $C=0$、1 和–1 的情况。数值结果发现，$C<0.5$ 时，指数脉冲时域波形峰值随传输距离的增加而单调减小，随啁啾 C 的增加而增加；$C\geqslant 0.5$ 时，其峰值首先经历一个初始增加阶段，后随传输距离的增加而减小，随啁啾 C 的增加而增加；啁啾 C 越大，峰值在 $\xi=1$ 附近的初始增加越大。

3.3.3　初始非线性啁啾对脉冲时域波形的影响

数值研究指数脉冲线性传输时域波形变化过程中，当指数脉冲边缘加入式(3-23)所示非线性啁啾($\tau\leqslant -2.1875$, $\phi=a$; $\tau\geqslant 2.1875$, $\phi=b$; $|\tau|<2.1875$, $\phi=-C\tau^2/2$; a 和 b 都是常数，$|\tau|=7\times 0.3125=2.1875$ 对应脉冲的边缘)微扰时，观察到了指数脉冲在线性传输中的波形分裂现象。

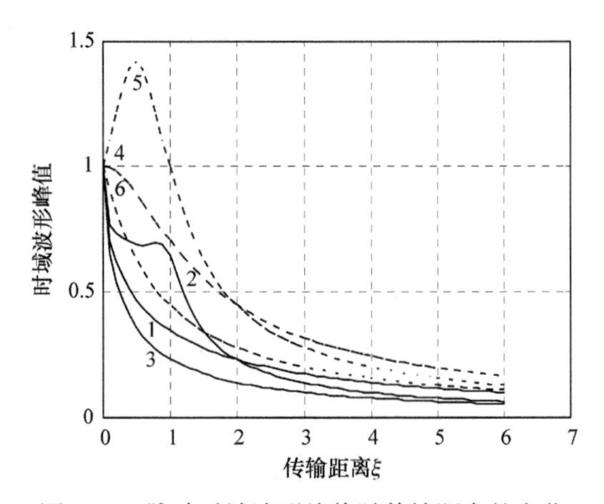

图 3-19　脉冲时域波形峰值随传输距离的变化

实线 1～3 所示对应指数脉冲 $C=0$、1 和–1 的情况，点线 4～6 所示对应高斯脉冲 $C=0$、1 和–1 的情况

图 3-20 所示是当式(3-23)中 $a=b=3$ 和 $C=-1$ 时指数脉冲线性传输到 $\xi=6$ 处的时域波形。本图与图 3-11 坐标、图例相同。数值结果表明，当 $a=b$ 时，出现对称的时域波形分裂现象，其前峰和后峰一样高，如图 3-20 实线所示；当 $a>b$ 时，指数脉冲在线性传输中出现了不对称的时域波形分裂现象，其中前峰比后峰低；当 $a<b$ 时，出现另一种不对称的时域波形分裂现象，其中前峰比后峰高。啁啾参量$|C|$值越大，波形分裂和波形边缘的色散波越强。

图 3-21 所示是当式(3-23)中 $a=b=3$ 和 $C=-1$ 时非线性啁啾指数脉冲的频谱。图中横坐标是归一化频率，纵坐标是频谱强度，任意单位。由图 3-21 可得，非线性啁啾改变了指数脉冲的初始频谱，使初始频谱分裂为双峰结构；脉冲在光纤中线性传输时，两个谱峰传输速度不同导致脉冲的时域波形分裂。啁啾参量$|C|$越大，脉冲频谱变化越剧烈，导致时域波形的分裂越严重。

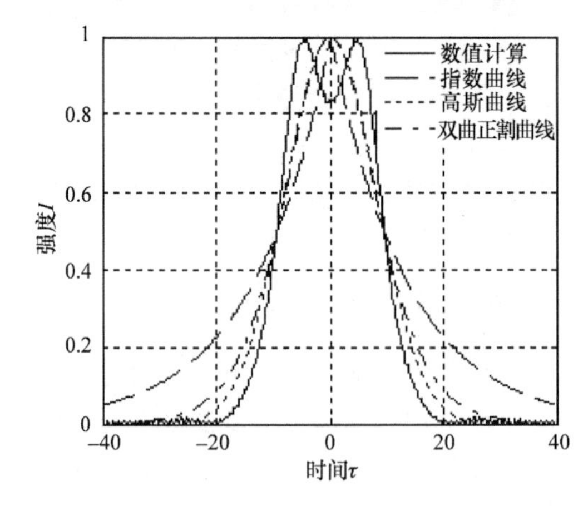

图 3-20　式(3-23)所示初始非线性啁啾指数脉冲在 $a=b=3$、$C=-1$ 和 $\xi=6$ 时的时域波形

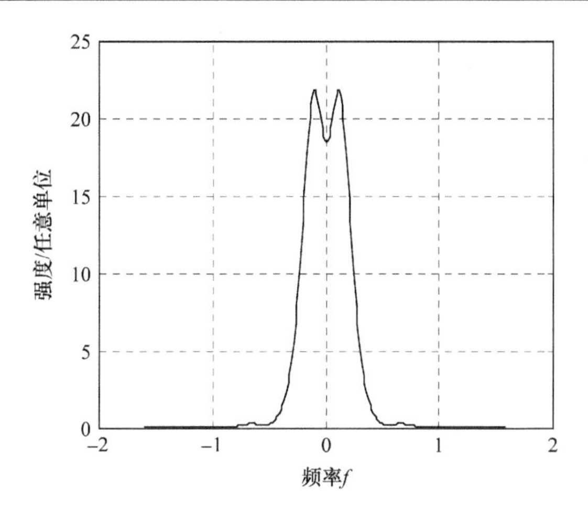

图 3-21　式(3-23)所示初始非线性啁啾指数脉冲在 $a=b=3$、$C=-1$ 时的频谱

3.4　本章小结

本章完成了 10GHz 短脉冲在色散平坦光纤中的线性传输实验，利用能够准确测量脉冲时域波形等特性的 SHG-FROG 技术对短脉冲线性传输进行了实验研究，并与高斯脉冲线性传输理论预期和数值计算结果进行了比较[90]。结果表明，激光器输出的短脉冲是具有负线性啁啾的高斯脉冲；经过 12.7km 色散平坦光纤线性传输后仍然为具有负线性啁啾的高斯脉冲，其谱宽在传输过程中基本保持不变，脉宽展宽了 3.1 倍，啁啾增大了 4 倍。实验测量结果、数值计算结果和高斯脉冲的线性传输理论预期一致，表明 SHG-FROG 脉冲分析仪实用性强，是探测、分析短脉冲的有效工具。

在此基础上，数值研究了初始啁啾双曲正割脉冲和指数脉冲的线性传输特性，填补了人们对其线性传输规律认识的国际空白，给出了两脉冲频谱宽度和时间带宽积随频率啁啾变化的表达式，提供了判断脉冲时域波形的一种有效方法[91,92]。结果表明，负线性啁啾对双曲正割和指数脉冲时域宽度和时间带宽积的变化影响比正啁啾的影响要大得多。非线性啁啾对双曲正割脉冲和指数脉冲线性传输时域波形变化的影响比线性啁啾更大，非线性啁啾指数脉冲和双曲正割脉冲在线性传输过程中都出现时域波形分裂现象，比具有相同啁啾的高斯脉冲时域波形分裂都严重。

参考文献

[1] 阿戈沃. 非线性光纤光学原理及应用.贾东方, 余震虹, 谈斌, 译. 北京: 电子工业出版社, 2002.

[2] Mollenauer L F, Stolen R H, Gordon J P. Experimental observation of picosecond pulse narrowing and solitons in optical fiber. Phys. Rev. Lett., 1980, 45(13): 1095～1098.
[3] Gouveia-Neto A S, Wigley P G J , Taylor J R. Soliton generation through Raman amplification of noise bursts. Optics Letters, 1989, 14(20): 1122～1124.
[4] Iwatsuki K, Suzuki K, Nishi S. Adiabatic soliton compression of gain-switched DFB-LD pulse by distributed fiber Raman amplification. IEEE Transactions Photonics Technology Letters, 1991, 3(12): 1074～1076.
[5] Murphy T E. 10-GHz 1.3-ps pulse generation using chirped soliton compression in a Raman gain medium. IEEE Photonics Technology Letters, 2002, 14(10): 1424～1426.
[6] Mollenauer L F, Stolen R H , Islam M N. Experimental demonstration of soliton propagation in long fibers: loss compensated by Raman gain. Optics Letters, 1985, 10(5): 229～231.
[7] Iwatsuki K, Nishi S, Saruwatari M, et al. 5Gb/s optical soliton transmission experiment using Raman amplification for fiber-loss compensation. IEEE Photonics Technology Letters, 1990, 2(7): 507～509.
[8] Okhrimchuk A G, Onishchukov G, Lederer F. Long-haul soliton transmission at 1.3 μm using distributed Raman amplification. Journal of Lightwave Technology, 2001, 19(6): 837～841.
[9] 高以智, 姚敏玉, 许宝西, 等. 2.5GHz 光孤子传输. 高技术通讯, 1994, 7: 4～6.
[10] 许宝西, 李京辉, 姜新, 等. 2.5GHz 光孤子传输. 电子学报, 1995, 23(11): 38～54.
[11] 杨祥林, 毛庆和, 温扬敬, 等. 30km 2.5GHz 光孤子波传输与压缩实验研究. 高技术通讯, 1996, 10: 26～28.
[12] 余建军, 杨伯君, 余建国, 等. 光孤子传输实验研究. 光电子 · 激光, 1996, 7(5): 267～272.
[13] 余建军, 杨伯君, 管克俭. 5GHz 的 16.2ps 超短光脉冲的产生. 光学学报, 1998, 18(1) :14～17.
[14] 余建军, 杨伯君, 管克俭. 基于不同色散光纤的光纤链的孤子传输研究. 光学学报, 1998, 18(4): 446～450.
[15] 张晓光, 林宁, 张涛, 等. 预啁啾 10GHz, 38km 色散管理孤子的传输实验. 光子学报, 2001, 30(7): 813～817.
[16] Matos C J S, Talor J R. Tunable repetition-rate multiplication of a 10 GHz pulse train using linear and nonlinear fiber propagation. Applied Physics Letters, 2003, 83(26): 5356～5358.
[17] 王安斌, 伍剑, 拱伟, 等. 高消光比超短脉冲产生的实验研究. 中国激光, 2004, 31(3): 265～268.
[18] 王兴涛, 印定军, 帅斌, 等. 应用全反射二阶自相关仪测量超短脉冲脉宽. 中国激光, 2004, 31(8): 1018～1020.
[19] Lin Q, Wright K, Agrawal G P, et al. Spectral responsitivity and efficiency of metal-based femtosecond autocorrelation technique. Optics Communications, 2004, 242: 279～283.
[20] Dai J M, Teng H, Guo C L. Second- and third-order interferometric autocorrelations based on harmonic generations from metal surfaces. Optics Communications, 2005, 252: 173～178.
[21] Fittinghoff D N, Au J A D, Squier J. Spatial and temporal characterizations of femtosecond pulses at high-numerical aperture using collinear, background-free, third-harmonic autocorrelation. Optics Communications, 2005, 247: 405～426.
[22] Wang S, Wang Y B, Feng G Y, et al. Generation of double-scale pulses in a LD-pumped

Yb:phosphate solid-state laser. Appl. Opt., 2017, 56: 897～900 .

[23] Marec A L, Guilbaud O, Larroche O, et al. Evidence of partial temporal coherence effects in the linear autocorrelation of extreme ultraviolet laser pulses. Opt. Lett., 2016, 41: 3387～3390.

[24] Chaparro A, Furfaro L, Balle S. Subpicosecond pulses in a self-starting mode-locked semiconductor-based figure-of-eight fiber laser. Photon. Res., 2017, 5: 37～40.

[25] Lauterio-Cruz P, Hernandez-Garcia J C, Pottiez O. High energy noise-like pulsing in a double-clad Er/Yb figure-of-eight fiber laser. Opt. Express, 2016, 24: 13778～13787 .

[26] Lin J H, Chen C L, Chan C W, et al. Investigation of noise-like pulses from a net normal Yb-doped fiber laser based on a nonlinear polarization rotation mechanism. Opt. Lett., 2016, 41: 5310～5313.

[27] Zhang F, Fan X W, Liu J,et al. Dual-wavelength mode-locked operation on a novel Nd^{3+}, Gd^{3+}: SrF_2 crystal laser. Opt. Mater. Express, 2016, 6: 1513～1519 .

[28] Chao M S, Cheng H N, Fong B J, et al. High-sensitivity ultrashort mid-infrared pulse characterization by modified interferometric field autocorrelation. Opt. Lett., 2015, 40: 902～905.

[29] Sun B S, Salter P S, Booth M J. Pulse front adaptive optics: A new method for control of ultrashort laser pulses. Opt. Express, 2015, 23: 19348～19357.

[30] Lin S S, Hwang S K, Liu J M. High-power noise-like pulse generation using a 1.56-μm all-fiber laser system. Opt. Express, 2015, 23: 18256～18268.

[31] Traore A, Lalanne E, Johnson A M. Determination of the nonlinear refractive index of multimode silica fiber with a dual-line ultra-short pulse laser source by using the induced grating autocorrelation technique. Opt. Express, 2015, 23: 17127～17137.

[32] Suzuki M, Ganeev R A, Yoneya S, et al. Generation of broadband noise-like pulse from Yb-doped fiber laser ring cavity. Opt. Lett.,2015, 40: 804～807.

[33] Tian W L, Wang Z H, Liu J X, et al. Dissipative soliton and synchronously dual-wavelength mode-locking Yb:YSO lasers. Opt. Express, 2015, 23: 8731～8739.

[34] Kane D J, Trebino R. Characterization of arbitrary femtosecond pulses using frequency-resolved optical gating. IEEE J. Quantum Electron., 1993, 29(2): 571～579.

[35] DeLong K W, Trebino R, Hunter J, et al. Frequency-resolved optical gating with the use of second-harmonic generation. J. Opt. Soc. Am. B,1994, 11(11): 2206～2215.

[36] Gallmann L, Steinmeyer G, Sutter D H, et al. Collinear type II second-harmonic-generation frequency-resolved optical gating for the characterization of sub-10-fs optical pulses. Optics Letters, 2000, 25(4): 269～271.

[37] 王兆华, 魏志义, 滕浩, 等. 飞秒激光脉冲的谐波频率分辨光学开关法测量研究. 物理学报, 2003, 52(2): 362～366.

[38] 龙井华, 高继华, 巨养锋, 等. 用 SHG-FROG 方法测量超短光脉冲的振幅和相位. 光子学报, 2002, 31(10): 1292～1296.

[39] DeLong K W, Fittinghoff D N, Trebino R, et al. Pulse retrieval in frequency-resolved optical gating based on the method of generalized projections. Optics Letters, 1994, 19(24): 2152～2154.

[40] Hu J, Zhang G Z, Zhang B G, et al. Using frequency-resolved optical gating to retrieve amplitude and phase of ultrashort laser pulse. Journal of Optoeletronics · Laser, 2002, 13(3): 232～236.

[41] Lacourt P A, Dudley J M, Merolla J M, et al. Milliwatt -peak-power pulse characterization at 1.55 μm by wavelength-conversion frequency-resolved optical gating. Optics Letters, 2002, 27(10): 863～865.

[42] Barry L P, Delburgo S, Thomsen B C, et al. Optimization of optical data transmitters for 40-Gb/s lightwave systems using frequecy resolved optical gating.IEEE Photon. Tech. Lett., 2002, 14(7): 971～973.

[43] Liu S, Lu D, Zhao L, et al. SHG-FROG characterization of a novel multichannel synchronized AWG-based mode-locked laser //Conference on Lasers and Electro-Optics, OSA Technical Digest (online) (Optical Society of America, 2017), paper JTh2A.131.

[44] Kane D J. Improved principal components generalized projections algorithm for frequency resolved optical gating //Conference on Lasers and Electro-Optics, OSA Technical Digest (online) (Optical Society of America, 2017), paper STu3I.4.

[45] Hyyti J, Escoto E, Steinmeyer G, et al. Interferometric time-domain ptychography for ultrafast pulse characterization. Opt. Lett., 2017, 42: 2185～2188.

[46] Sidorenko, Lahav O, Avnat Z, et al. Ptychographic reconstruction algorithm for frequency-resolved optical gating: Super-resolution and supreme robustness. Optica, 2016, 3: 1320～1330.

[47] Heidt A M, Spangenberg D M, Brügmann M, et al. Improved retrieval of complex supercontinuum pulses from XFROG traces using a ptychographic algorithm. Opt. Lett., 2016, 41: 4903～4906.

[48] Ermolov A, Valtna-Lukner H, Travers J, et al. Characterization of few-fs deep-UV dispersive waves by ultra-broadband transient-grating XFROG. Opt. Lett., 2016, 41: 5535～5538.

[49] Fuji T, Shirai H, Nomura Y. Self-referenced frequency-resolved optical gating capable of carrier-envelope phase determination //Conference on Lasers and Electro-Optics, OSA Technical Digest (2016) (Optical Society of America, 2016), paper SM3I.7.

[50] Steinmeyer A. Interferometric FROG for Ultrafast Spectroscopy on the Few-cycle Scale //Conference on Lasers and Electro-Optics, OSA Technical Digest (2016) (Optical Society of America, 2016), paper STu4I.1.

[51] Okamura B, Sakakibara Y, Omoda E, et al. Experimental analysis of coherent supercontinuum generation and ultrashort pulse generation using cross-correlation frequency resolved optical gating (X-FROG). J. Opt. Soc. Am. B, 2015, 32: 400～406 .

[52] Itakura R, Kumada T, Nakano M, et al. Frequency-resolved optical gating for characterization of VUV pulses using ultrafast plasma mirror switching. Opt. Express, 2015, 23: 10914～10924.

[53] Snedden E W, Walsh D A, Jamison S P. Revealing carrier-envelope phase through frequency mixing and interference in frequency resolved optical gating. Opt. Express, 2015, 23: 8507～8518.

[54] Hause A, Kraft S, Rohrmann P, et al. Reliable multiple-pulse reconstruction from second-harmonic-generation frequency-resolved optical gating spectrograms. J. Opt. Soc. Am. B, 2015, 32: 868～877.

[55] Li X J, Liao J L, Nie Y M, et al. Unambiguous demonstration of soliton evolution in slow-light silicon photonic crystal waveguides with SFG-XFROG. Opt. Express, 2015, 23: 10282～10292.

[56] Hasegawa A, Tappert F. Transmission of stationary nonlinear optical pulses in dispersive dielectric fibers. Ⅰ. Anomalous dispersion. Appl. Phys. Lett., 1973, 23(3): 142～144.

[57] 刘颂豪, 赫光生. 强光光学及其应用. 广州: 广东科技出版社, 1995.
[58] 庞小峰. 孤子物理学. 北京: 科学出版社, 1987.
[59] 杨祥林, 温扬敬. 光纤孤子通信理论基础. 北京: 国防工业出版社, 2000.
[60] 黄景宁, 徐济仲, 熊吟涛. 孤子概念、原理和应用. 北京: 高等教育出版社, 2004.
[61] Ablowitz M J, Kaup D J, Newell A C, et al. The inverse scattering transform-fourier analysis for nonlinear problems. Studies in Applied Mathematics, 1974, 53(4): 249～315.
[62] Mollenauer L F, Lichtman E, Harvey G T, et al. Demonstration of error-free soliton transmission over more than 15000km at 5Gbit/s single-channel, and over 11000km at l0Gbit/s in two-channel WDM. Electronics Letters, 1992, 28(8): 792～794.
[63] Favre F, Guen D L, Devaux F. 4×20Gbit/s soliton WDM transmission over 2000km with 100km dispersion compensated spans of standard fibre. Electronics Letters, 1997, 33 (14): 1234～1235.
[64] Favre F, Guen D Le, Moulinard M L, et al. 16×20Gbit/s soliton WDM transmission over 1300 km with 100 km dispersion compensated spans of standard fibre. Electronics Letters, 1997, 33 (25): 2135～2136.
[65] Sahara A, Kubota H, Nakazawa M. Experiment and analyses of 20-Gbit/s soliton transmission systems using installed optical fiber cable. Electronics and Communications in Japan, Part 2, 1998, 81(1): 74～83.
[66] Nakazawa M, Suzuki K, Kubota H. 160Gbit/s (80Gb/s×2channels) WDM soliton transmission over 10000km using in-line synchronous modulation. Electronics Letters, 1999, 35(16): 1358～1359.
[67] Takushima Y, Douke T, Wang X M, et al. Dispersion tolerance and transmission distance 1000km of a 40-Gb/s dispersion management soliton transmission system. Journal of Lightwave Technology, 2002, 20(3): 360～367.
[68] Gouveia-Neto A S, Wigley P G J, Taylor J R. Soliton generation through Raman amplification of noise bursts. Optics Letters, 1989, 14(20): 1122～1124.
[69] Iwatsuki K, Suzuki K, Nishi S. Adiabatic soliton compression of gain-switched DFB-LD pulse by distributed fiber Raman amplification. IEEE Transactions Photonics Technology Letters, 1991, 3(12): 1074～1076.
[70] Murphy T E. 10-GHz 1.3-ps pulse generation using chirped soliton compression in a Raman gain medium. IEEE Photonics Technology Letters, 2002, 14(10): 1424～1426.
[71] Mollenauer L F, Stolen R H, Islam M N. Experimental demonstration of soliton propagation in long fibers: Loss compensated by Raman gain. Optics Letters, 1985, 10(5): 229～231.
[72] Iwatsuki K, Nishi S, Saruwatari M, et al. 5Gb/s optical soliton transmission experiment using Raman amplification for fiber-loss compensation. IEEE Photonics Technology Letters, 1990, 2(7): 507～509.
[73] Okhrimchuk A G, Onishchukov G, Lederer F. Long-haul soliton transmission at 1.3 μm using distributed Raman amplification. Journal of Lightwave Technology, 2001, 19(6): 837～841.
[74] Ereifej H N, Grigoryan V, Carter G M. 40 Gbit/s long-haul transmission in dispersion-managed soliton system using Raman amplification. Electronics Letters, 2001, 37(25): 1538～1539.
[75] Pincemin E, Hamoir D, Audouin O, et al. Distributed-Raman-amplification effect on pulse

interactions and collisions in long-haul dispersion-managed soliton transmissions. J. Opt. Soc. Am. B, 2002, 19(5): 973～980.
[76] Tio A A B, Shum P. Propagation of optical soliton in a fiber Raman amplifier. Proceedings of SPIE, 2004, 5280: 676～681.
[77] Mollenauer L F, Smith K. Demonstration of soliton transmission over more than 4000km in fiber with loss periodically compensated by Raman gain. Optics Letters, 1988, 13(8): 675～677.
[78] Chi S, Wen S. Interaction of optical solitons with a forward Raman pump wave. Optics Letters, 1989, 14(1): 84～86.
[79] Wen S, Wang T Y, Chi S. The optical soliton transmission amplified by bidirectional Raman pumps with nonconstant depletion. IEEE Journal of Quantum Electronics, 1991, 21(8): 2066～2073.
[80] Levy G F. Raman amplification of solitons in a fiber optic ring. Journal of Lightwave Technology, 1996, 14(1): 72～76.
[81] Liu S L, Wang W Z, Xu J Z. Exact N-soliton solutions of the modified nonlinear Schrödinger equation. Phys. Rev. E, 1994, 49 (6): 5726～5730.
[82] Liu S L, Wang W Z. Complete compensation for the soliton self-frequency shift and third-order dispersion of a fiber. Opt. Lett., 1993,18(22): 1911～1912.
[83] Liu S L, Liu X Q. Mutual compensation of the higher-order nonlinearity and the third-order dispersion, Phys. Lett. A, 1997, 225(1-3): 67～72.
[84] 曹文华, 刘颂豪, 廖常俊, 等. 色散缓变光纤中的孤子效应拉曼脉冲产生. 中国激光, 1994, 21(6): 489～494.
[85] 李宏, 杨祥林, 刘堂坤. 暗孤子传输系统中调制拉曼泵浦的控制作用. 中国激光, 1997, 24(7): 654～658.
[86] 沈廷根, 郑浩, 李正华, 等. 掺杂光子晶体光纤的缺陷模增益谱与光孤子拉曼放大研究. 人工晶体学报, 2005, 34(6): 1065～1073.
[87] MacLeod A M, Yan X, Gillespie W A, et al. Formation of low time-bandwidth product, single-sided exponential optical pulses in free-electron laser oscillators. Physical Review E, 2000, 62(3): 4216～4220.
[88] Shapiro S L.超短光脉冲-皮秒技术及其应用. 朱世清, 译. 北京: 科学出版社, 1987.
[89] 朱京平. 光电子技术基础. 成都: 四川科学技术出版社, 2003.
[90] 刘山亮, 郑宏军, 短脉冲在色散平坦光纤中传输前后波形、相位和啁啾测量的实验研究. 中国激光, 2006, 33(2): 199～205.
[91] 郑宏军, 刘山亮, 黎昕, 等. 初始啁啾对双曲正割光脉冲线性传输特性的影响. 物理学报, 2007, 56(4): 2286～2292 .
[92] Zheng H J, Liu S L. Effects of initial frequency chirp on the linear propagation characteristics of the exponential optical pulse. Chinese Physics, 2006, 15(08): 1831～1837.

第 4 章　啁啾脉冲在标准单模光纤通信系统中的非线性传输研究

非线性传输问题，特别是由非线性薛定谔方程所描述的孤子问题出现在现代科学的各个分支中，例如，信息科学中的光纤孤子等都可以用非线性薛定谔方程描述。由于光孤子广阔、明朗的应用前景和易于实验研究等特点，几十年来得到了广泛的研究和发展，为物理学、等离子体、信息科学、生命科学和其他学科中众多的类似问题的解决做出了重大贡献[1-5]。鉴于以往实验条件等各种因素的限制[1, 6-31]，本章利用能够准确测量脉冲时域波形等特性的 SHG-FROG 技术研究了脉冲在单模光纤中传输前后脉冲宽度、波形、啁啾和时间带宽积等的变化以及脉冲演化形成孤子的规律和特点，并与理论预期作了对比。考虑到以往对指数脉冲的研究仅限于 $T \geqslant 0$ 的情况[32-34]，双边指数脉冲($-\infty<T<\infty$)的非线性传输特性难于解析研究，在研究脉冲形成孤子的实验和理论基础上，数值研究了啁啾双边指数脉冲的非线性传输特性。本章研究为光孤子光源及其通信系统的设计优化提供了重要依据。

4.1　啁啾脉冲演化成孤子的实验研究

4.1.1　实验装置

图 4-1 所示是利用 SHG-FROG 脉冲分析仪测量短脉冲非线性传输特性的实验装置。实验所用激光器为德国 U2T 公司的可调谐半导体锁模脉冲激光器 TMLL1550，光隔离器为武汉光讯公司生产，EDFA 为法国 KPS 公司的掺铒光纤放大器，FROG 分析仪为新西兰南方光子有限公司的 SHG-FROG 脉冲分析仪 HR200，实验装置中仪器设备的详细说明见 3.1 节。在本节所述实验中半导体锁模激光器输出的脉冲经光隔离器进入掺铒光纤放大器放大后输入光纤，光纤输出

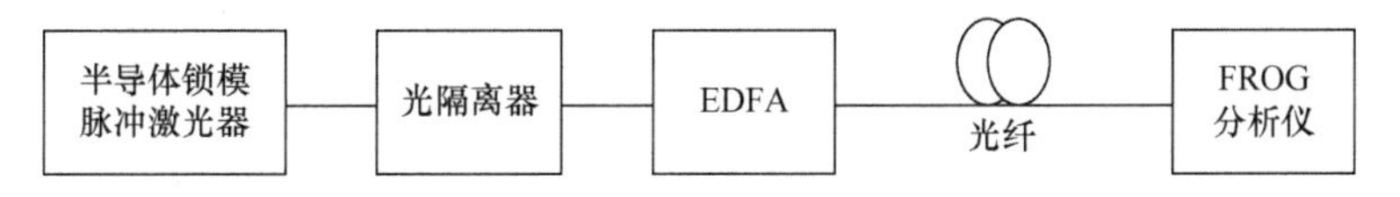

图 4-1　脉冲非线性传输实验装置

脉冲经偏振控制器(PC)输入分析仪 HR200 进行测量和分析。各器件之间用光纤活动连接器或跳线连接。实验中所用光纤为 G.652 标准单模光纤，其模场直径为 9.07μm、色散参量 D=15.07ps/(nm · km)，光纤损耗为 0.188dB/km，长度分别是 200m、355m 和 1000m。

4.1.2 实验结果与分析

在接入光纤前，作者测量分析了输入脉冲特性。图 4-2(a)所示是测量得到的 10GHz、1550nm 脉冲时域波形。图中虚线是高斯曲线，点划线是双曲正割曲线，实线是实验测量得到的时域波形，图 4-2(a)中横坐标是时间，单位皮秒(ps)，纵坐标是脉冲的归一化强度(intensity)，下文中图 4-4 和图 4-5 中坐标、图例与此相同。由图 4-2(a)可见，输入脉冲前沿下降较高斯曲线下降慢，而后沿下降较高斯曲线下降稍快。测量结果表明输入脉冲的半峰全宽为 1.86ps、谱宽为 2.22nm，时间带宽积为 0.514。

图 4-2(b)所示是输入脉冲的啁啾曲线，横坐标是时间 T 与半峰全宽 T_{FWHM} 的比值，纵坐标表示脉冲频率啁啾 $\delta\omega/\omega$。本节图 4-6～图 4-8 与本图坐标相同。这里 $\delta\omega=-\dfrac{\partial\phi}{\partial T}$ 根据测量得到的脉冲相位曲线，即 ϕ-T 曲线给出。由图 4-2(b)可见，在脉冲中心附近啁啾是线性的，脉冲前沿的啁啾比后沿的稍大，这与脉冲时域波形前沿比后沿下降慢有关。

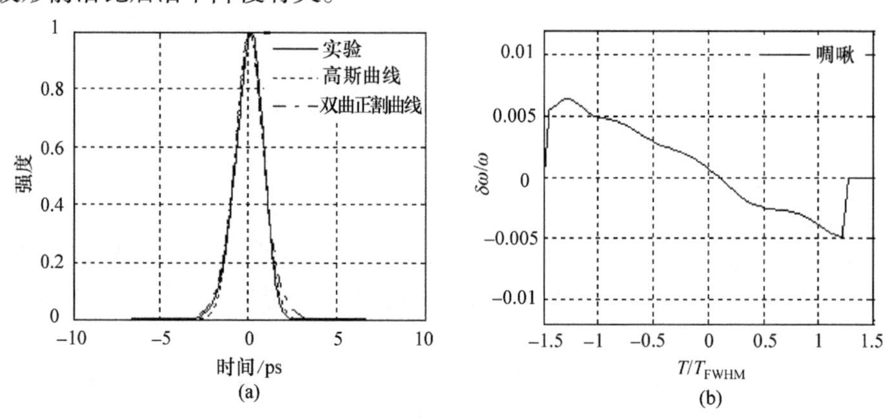

图 4-2 传输前测量得到的脉冲时域波形(a)和频率啁啾曲线(b)

4.1.2.1 脉冲时域宽度与输入功率的关系

在图 4-1 所示实验装置中，在 EDFA 和偏振控制器(polarization controller)之间先后接入 200m、355m 和 1000m 具有同样参数的 G.652 光纤。EDFA 输出的脉冲经光纤传输后输入 HR200 进行测量和分析。通过调节 EDFA 泵浦激光器的工作电流改变输入光纤脉冲的功率后，再进行测量和分析。图 4-3 所示是接入不同长度光纤时实验测量得到的输出脉冲时域半峰全宽随输入平均光功率的变化关系。图中横坐标是输入平均光功率，单位是毫瓦(mW)，纵坐标是脉冲时域半峰全宽，单位是皮秒(ps)。从图中看出，对于确定的光纤长度，输出脉冲的时域宽度随着输入脉冲平均功率的增加而减小。

当输入功率小于 240mW(23.8dBm)时，在相同输入功率下输出脉冲宽度随着传输距离而增加；在相同传输距离下输入功率越低，输出脉冲宽度越宽；输入功率越低，经过不同传输距离输出脉冲的宽度相差越大。这表明当输入功率小于 240mW 时，色散引起的脉冲展宽占主导地位。

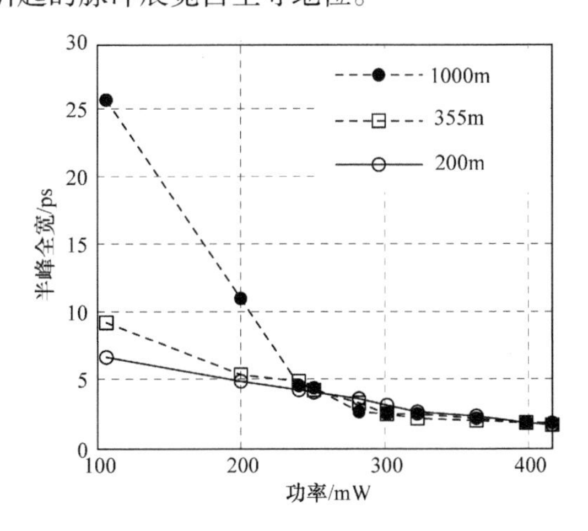

图 4-3　脉冲宽度随输入功率的变化

当输入功率大于等于 240mW 时，在相同输入功率下经过不同传输距离后的输出脉冲宽度基本相同，色散引起的脉冲展宽和非线性引起的脉冲压缩达到基本平衡，符合一阶孤子的基本特征。这表明输入脉冲在光纤中经过 200m 传输后已经基本演化形成孤子，在此后的传输过程中其宽度基本保持不变。虽然在相同输入功率下经过不同传输距离后的输出脉冲宽度近似相等，但是输出脉冲宽度随着输入功率的增加而缓慢减小。

4.1.2.2　脉冲波形的演化

图 4-4(a)所示是输入功率为 240mW 时输入脉冲在光纤中经过 200m 距离后输出脉冲的波形曲线。由图可见，实验测量得到的输出脉冲波形曲线与双曲正割曲线完全吻合。这表明输入脉冲在光纤中经过 200m 距离后已经演化成为具有双曲正割波形的光纤孤子脉冲。孤子脉冲宽度为 4.23ps，比输入功率 107mW 时输出脉冲宽度 6.7ps 窄，比输入脉冲宽得多。

图 4-4(b)、(c)所示分别是输入功率为 240mW 时输入脉冲在光纤中经过

355m 和 1000m 距离后的波形曲线。由图可见，实验测量得到的经过 355m 距离后输出脉冲波形曲线仍与双曲正割曲线基本吻合，脉冲半峰全宽 4.64ps，比输入功率 107mW 时输出脉冲宽度 9.22ps 窄。经过 1000m 距离后输出脉冲波形曲线的前、后沿下降比双曲正割曲线稍微缓慢，脉冲半峰全宽 4.53ps，比输入功率 107mW 时输出脉冲宽度 25.6ps 窄得多。这表明经过 200m 距离后演化形成的光纤孤子脉冲在随后的传输过程中其宽度基本保持不变，其波形在双曲正割波形附近变化。

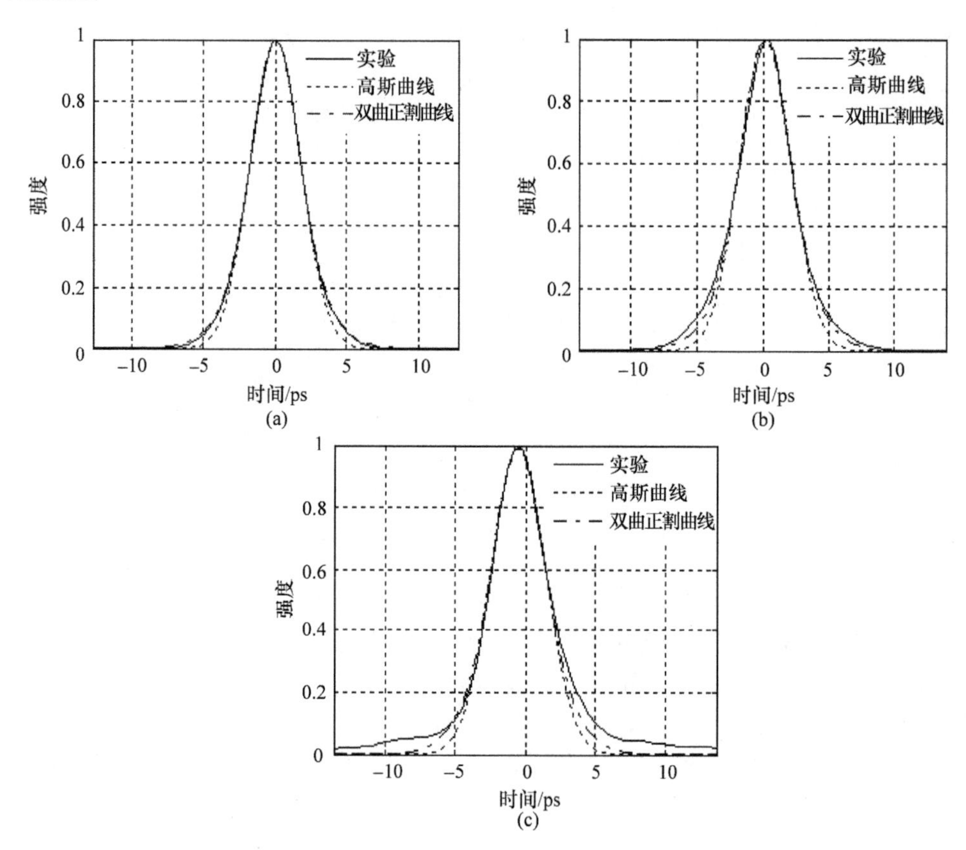

图 4-4　输入功率为 240mW 时输入脉冲在光纤中经过 200m(a)、355m(b)和 1000m(c)后输出脉冲的波形曲线

图 4-5(a)所示是输入功率 324mW(25.1dBm)时输入脉冲在光纤中经过 200m 距离后的波形曲线。由图可见，实验测量得到的输出脉冲波形曲线与双曲正割曲线完全吻合。这表明与输入功率 240mW 时一样，输入脉冲经过大约 200m 长度后已经演化成为具有双曲正割波形的光纤孤子脉冲。孤子脉冲宽度为 2.71ps，比

输入脉冲宽度稍宽，约为输入功率 240mW 时演化形成的孤子脉冲宽度的 2/3，大约是输入功率 107mW 时输出脉冲宽度 6.7ps 的 2/5。

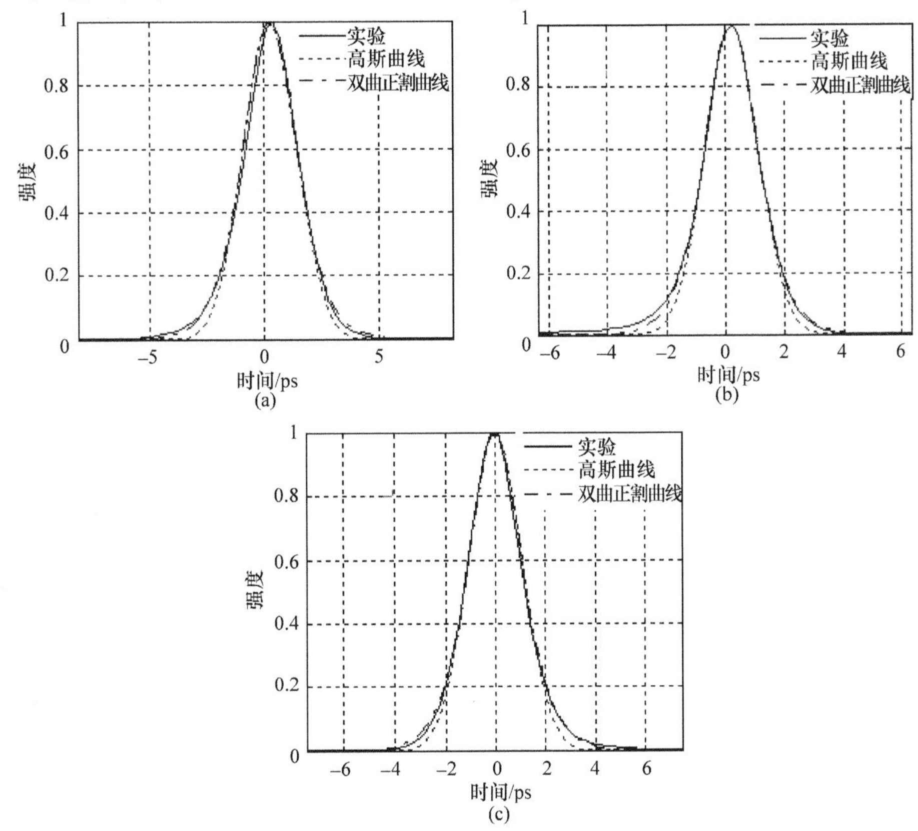

图 4-5　输入功率为 324mW 时光脉冲在光纤中传输 200m(a)，355m(b)和 1000m(c)后的时域波形曲线

图 4-5(b)所示是输入功率为 324mW 时输入脉冲在光纤中经过 355m 距离后的波形曲线。由图可见，实验测量得到的输出脉冲波形曲线接近双曲正割波形，输出脉冲宽度与经过 200m 距离后的宽度基本相同，是输入功率 107mW 时输出脉冲宽度 9.22ps 的 2/7。

图 4-5(c)所示是输入功率为 324mW 时输入脉冲在光纤中经过 1000m 距离后的波形曲线。由图可见，实验测量得到的输出脉冲波形曲线与双曲正割曲线完全吻合，与经过 200m 距离后输出脉冲的波形和宽度基本相同，输出脉冲宽度只是输入功率 107mW 时输出脉冲宽度 25.6ps 的百分之十。

4.1.2.3 脉冲啁啾的演化

当输入功率小于 240mW(23.8dBm)时，输出脉冲的啁啾是负线性的，其大小随着输入功率的增加而减小，色散引起的负线性啁啾占主导地位。在相同输入功率下，输出脉冲的啁啾随着传输距离的增加而减小。

图 4-6 所示是输入功率为 107mW(20.3dBm)时输入脉冲在光纤中传输 200m、355m 和 1000m 后的啁啾曲线。由图可见，输出脉冲中心区域的啁啾曲线近似为一条斜率为负的直线，斜率的绝对值和啁啾的大小随着传输距离的增大而减小。这表明非线性的影响随着传输距离的增加而增加，但是色散的影响仍占主导地位。

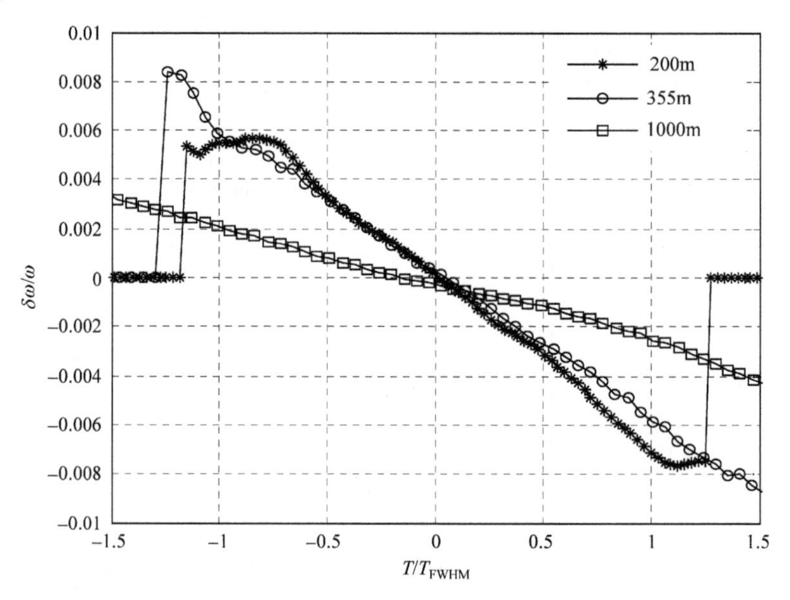

图 4-6 输入功率 107mW 时脉冲在光纤中传输 200m，355m 和 1000m 后的啁啾曲线

图 4-7 所示为输入功率 240mW 时输入脉冲在光纤中传输 200m、355m 和 1000m 后的啁啾曲线。

由图 4-7 可见，输出脉冲啁啾的大小随着传输距离的增大而减小。输入脉冲在光纤中传输 200m 后在 $T=-1\sim1T_{\mathrm{FWHM}}$ 的中心区域的啁啾曲线仍然近似为一条直线，但是斜率的绝对值和啁啾的大小比 107mW 时显著减小。输入脉冲在光纤中传输 355m 后在 $T=-1.5\sim0T_{\mathrm{FWHM}}$ 区域的啁啾曲线仍然近似为一条直线，但是斜率的绝对值和啁啾的大小比 107mW 时显著减小；在 T=0～$0.3T_{\mathrm{FWHM}}$ 区域的啁啾近似为零，在 $T>0.3T_{\mathrm{FWHM}}$ 区域的啁啾随着 T 增加而增加。在光纤中传输 1000m 后在 $T=-1.3\sim1.2T_{\mathrm{FWHM}}$ 区域的啁啾等于零。

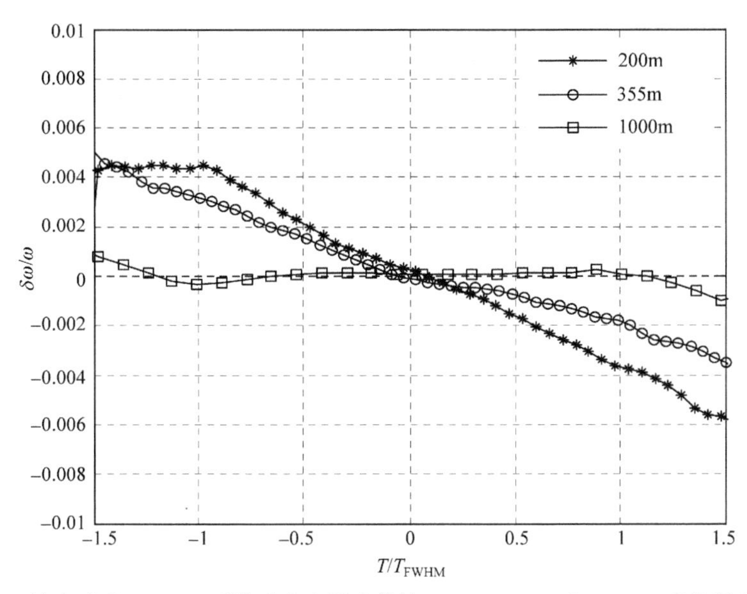

图4-7　输入功率240mW时脉冲在光纤中传输200m，355m和1000m后的啁啾曲线

图4-8所示输入功率为324mW时输入脉冲在光纤中传输200m、355m和1000m后的啁啾曲线。由图可见，输出脉冲中心附近的啁啾曲线近似为一条斜率为负的直线，传输355m和1000m后直线斜率的绝对值即啁啾的大小比传输200m后的小。

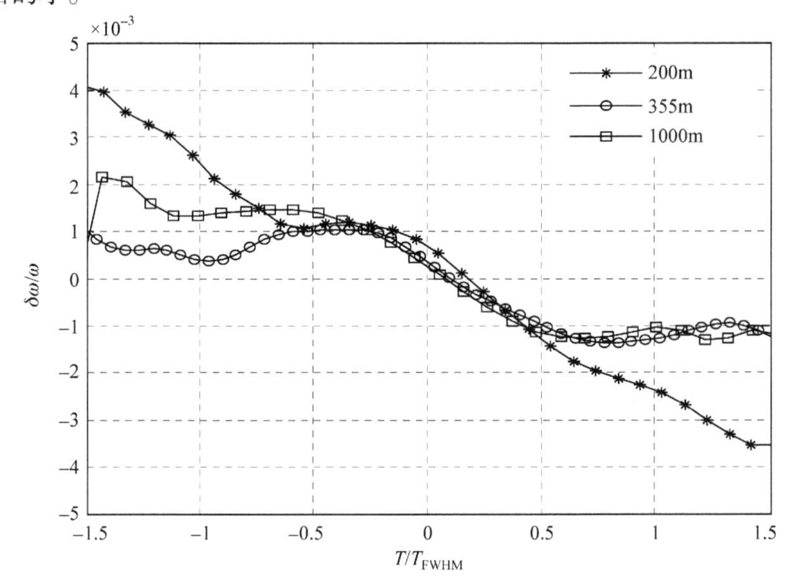

图4-8　输入功率324mW时光脉冲在光纤中传输200m，355m和1000m后的啁啾曲线

4.1.2.4 脉冲时间带宽积的演化

不仅脉冲啁啾随着传输距离和输入功率的变化而变化，而且其时间带宽积也随着传输距离和功率的变化而变化。实验发现，输入脉冲在光纤中传输 200m 时的时间带宽积随着输入功率的增加而减小，由输入功率 107mW 时的 1.264 减小到 324mW 时的 0.426。在随后的传输过程中脉冲的时间带宽积随着输入功率的变化而变化，输入功率 240mW 时经过 200m、355m 和 1000m 距离后的时间带宽积分别是 0.741、0.617 和 0.312，输入功率为 324mW 时分别是 0.475、0.323 和 0.367。

4.1.3 讨论

根据光纤孤子理论，在输入脉冲是无啁啾的双曲正割脉冲和没有光纤损耗的情况下，当输入功率等于基孤子峰值功率

$$P_0 = \frac{|\beta_2|}{\gamma T_0^2} \tag{4-1}$$

时，脉冲的波形和宽度在传输过程中保持不变。在以往的光纤孤子传输实验中都是以输出脉冲宽度等于输入脉冲宽度作为孤子形成的标志。根据上述孤子形成的标志，可以认为本节实验中输入脉冲在光纤中经过 200m 距离传输后已经演化形成孤子。在本书所述的实验条件下 $\beta_2 = -19.2\text{ps}^2/\text{km}$，$\gamma = 1.63\text{W}^{-1} \cdot \text{km}^{-1}$(取非线性克尔系数 $n_2 = 2.6\times10^{-20}\text{m}^2/\text{W}$)[1]，对半峰全宽为 1.86ps 的双曲正割输入脉冲 T_0=1.06ps，色散长度 L_D=0.058km。根据式(4-1)计算得到一阶孤子峰值功率为 10.6W，在本实验条件下与平均功率 223mW 对应。根据孤子微扰理论[35]，具有如下形式

$$U(0,T) = (1+\varepsilon)\text{sech}T, \quad |\varepsilon| < 0.5 \tag{4-2}$$

输入脉冲在光纤中可以演化形成一阶孤子，即$-0.5< \varepsilon <0.5$(输入功率在 0.25～2.25)的无啁啾双曲正割脉冲都能够演化形成一阶孤子。在本节所述实验中，当输入功率大于孤子功率理论值 223mW，即 $\varepsilon >0$ 时能够演化形成孤子；当输入功率小于孤子功率，即 $\varepsilon <0$ 时不能够演化形成孤子。实验结果与微扰理论的差异应该与输入脉冲的非双曲正割波形和啁啾有关。

在本节所述实验中当输入功率大于一阶孤子功率的理论值 223mW 时，输入脉冲在光纤中经过 200m(大约 3 个色散长度)传输都能够演化形成一阶孤子，但是所形成孤子脉冲宽度随着输入功率的变化而变化，输入功率越大，孤子脉冲的宽度越小。当输入功率为 240mW 时，经过 200m 光纤后所形成孤子脉冲的线性啁啾参数 C 为-0.9，半峰全宽为 4.23ps，远大于输入脉冲宽度 1.86ps。当输入功

率等于 324mW 时，所形成孤子脉冲半峰全宽 2.71ps 比输入功率 240mW 时的窄得多，其啁啾比输入功率 240mW 时的小得多。

4.2　啁啾孤子演化和传输的数值分析

4.2.1　输入脉冲特性

进一步研究表明，4.1 节实验的输入脉冲时域波形和相位随输入功率的增大而变化。在 240mW(23.8dBm) $\leqslant P_{\text{in}} \leqslant$ 324mW (25.1dBm)时，输入脉冲近似为线性啁啾高斯脉冲。当输入光功率在 240mW(23.8dBm)和 324mW(25.1dBm)之间变化时，将不同光功率的输入脉冲波形和相位数据由 SHG-FROG 脉冲分析仪导入 Matlab 计算程序中进行曲线拟合，将得到实验中输入脉冲的包络电场表达式和啁啾等参量。图 4-9 所示是在 P_{in}=240mW 时，输入脉冲的时域波形(a)和相位曲线(b)，图中横坐标是时间，单位是皮秒(ps)，图(a)纵坐标是脉冲归一化强度，图(b)纵坐标是脉冲相位，单位是弧度(rad)，实线所示是实验曲线，与线性啁啾高斯脉冲

$$u(0,\tau) = A\exp(-0.5\tau^2)\exp(-0.5\mathrm{i}C\tau^2) \tag{4-3}$$

的波形和相位曲线非常吻合。式(4-3)中，u 是包络电场，A=1 是振幅，$\tau = T/T_0$ 是归一化时间，T_0=1.799/1.665ps 是脉冲半宽度，$C = -0.6$ 是线性啁啾参量。

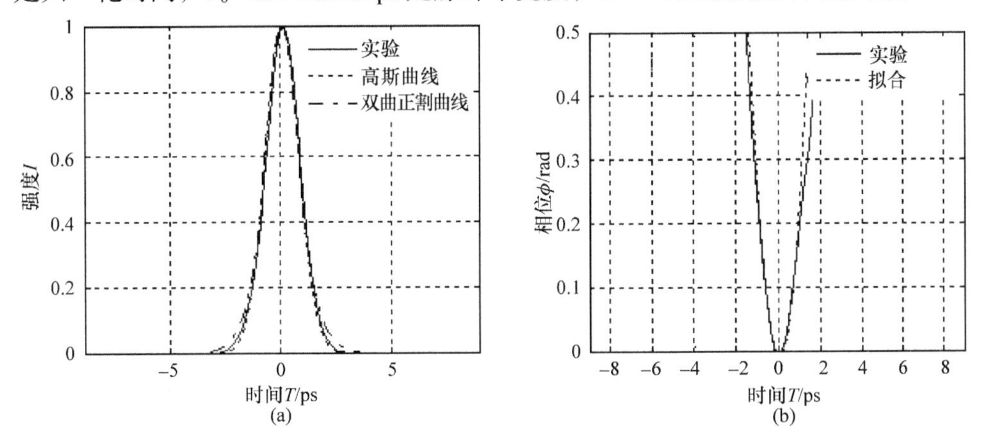

图 4-9　在 P_{in}=240mW 时输入脉冲时域波形(a)和相位曲线(b)

实线是实验曲线，点线是方程(4-3)的波形(a)和相位(b)曲线，点划线是双曲正割曲线

图 4-10 所示是在 P_{in}=324mW(25.1dBm)时，输入脉冲的时域波形(a)和相位曲线(b)，本图与图 4-9 相应坐标相同，实线所示是实验曲线，与式(4-3)所示包络电场吻合。其中，T_0=1.742/1.665ps 是脉冲半宽度，$C = -0.6$ 是线性啁啾参量。

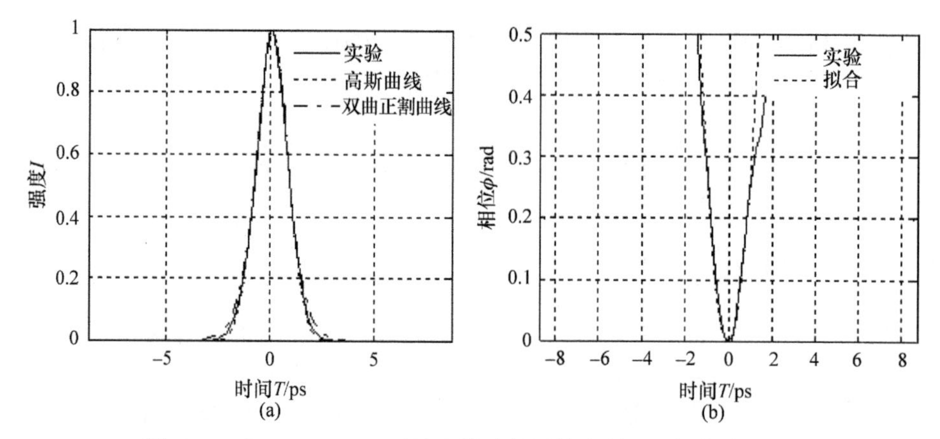

图 4-10　在 P_{in}=324mW 时输入脉冲的时域波形(a)和相位曲线(b)

实线是实验曲线，点线是方程(4-3)的波形(a)和相位(b)曲线，点划线是双曲正割曲线

4.2.2　啁啾孤子的演化形成

10GHz 短脉冲在 200m 光纤传输实验过程中，调节输入平均光功率在 240mW 和 324mW 之间变化，利用 SHG-FROG 脉冲分析仪测量研究输出脉冲的时域波形和相位曲线等特性，发现脉冲经过 200m 光纤传输后，其时域波形都与双曲正割脉冲波形一致。然后，作者以方程(4-3)为输入脉冲，采用分步傅里叶方法，按照非线性传输方程对脉冲在光纤中的传输进行了数值研究。考虑到传输脉冲是皮秒脉冲且传输距离较短，忽略光纤损耗、高阶色散项和高阶非线性项，广义非线性薛定谔方程(1-21)修正为

$$\mathrm{i}\frac{\partial u}{\partial \xi}+\frac{1}{2}\frac{\partial^2 u}{\partial \tau^2}+|u|^2 u=0 \tag{4-4}$$

式中，ξ 是传输距离，归一化到色散长度。$\tau=T/T_0$ 为归一化时间。平均输入光功率 240mW 和 324mW 分别归一化为 A=0.94 和 1.10。在 P_{in}=240mW 时，传输距离 200m、355m 和 1000m 分别归一化为 3.29、5.84 和 16.45 个色散长度。在 P_{in}=324mW 时，200m、355m 和 1000m 分别归一化为 3.51、6.23 和 17.55 个色散长度。

图 4-11 所示是 A=0.94(a)时在 200m 处的脉冲时域波形。实线所示是实验波形，虚线所示是数值计算结果，点线所示是与实验脉冲具有相同半峰全宽的高斯脉冲波形，点划线所示是相应的双曲正割脉冲波形，本图与图 4-9(a)坐标相同。图 4-12 所示是 A=0.94(a)和 1.10(b)时在 200m 处的脉冲啁啾曲线。实线所示是实验曲线，点线所示是数值计算结果。本图与图 4-2(b)坐标相同。由图 4-11 得到，在 200m 处，数值计算得到的输出脉冲时域波形与实验脉冲波形基本吻合，除了数值波形边缘下降比双曲正割脉冲的稍慢，都接近双曲正割脉冲波形；脉冲宽度

随光功率(A)增加而减小。由图 4-12 得到，在 200m 处，数值计算得到的 A=0.94 时的输出脉冲啁啾与实验数据非常吻合。A=1.10 时的啁啾比 A=0.94 时的要小，其数值计算的啁啾与实验数据的差别比 A=0.94 时的稍大。在脉冲中心附近，啁啾曲线都接近负线性啁啾。

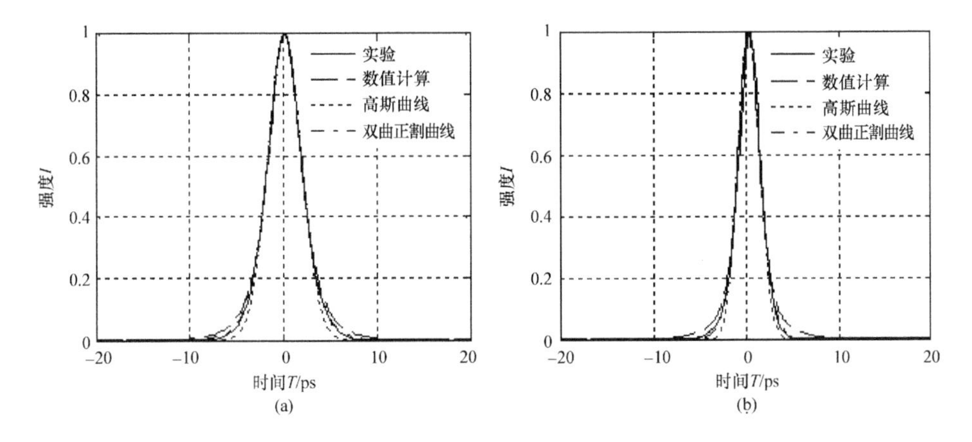

图 4-11　A=0.94(a)和 1.10(b)时在 200m 处的脉冲时域波形

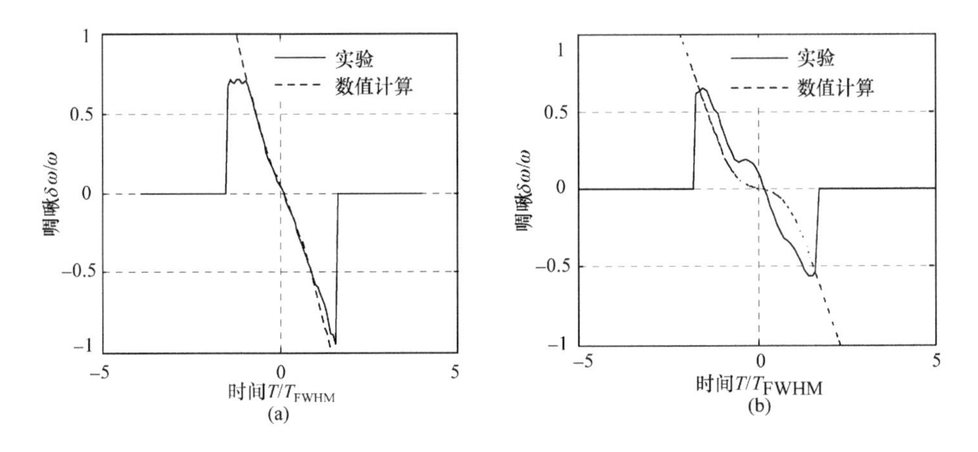

图 4-12　A=0.94(a)和 1.10(b)时在 200m 处的脉冲啁啾曲线

上述现象表明，输入脉冲在光纤中已经演化成啁啾孤子，孤子脉冲时域宽度和啁啾随脉冲振幅 A 的增加而减小，数值计算得到的脉冲时域波形、时域半峰全宽和啁啾与实验数据基本一致。双曲正割脉冲是方程(4-4)的解析解，是稳定的传输模式。然而，输入脉冲式(4-3)不是方程(4-4)的解析解，不能稳定传输，将演化为具有负啁啾的近双曲正割脉冲。

4.2.3 啁啾孤子的传输特性

作者数值研究了啁啾孤子的传输特性。图 4-13 所示是脉冲时域半峰全宽随传输距离的变化。实线所示是 A=0.94 时的数值结果，点线是 A=1.10 时的数值结果，小圆圈所示是 A=0.94 时的实验数据，小方框所示是 A=1.10 时的实验数据。由图 4-13 得到，脉冲的数值时域半峰全宽特性与 SHG-FROG 脉冲分析仪测量得到的实验数据基本吻合。脉冲时域半峰全宽随传输距离的增加产生衰减振荡，振荡振幅和周期随光功率(A)的增加而减小。脉冲传输过程中，脉冲压缩和展宽交替进行。负啁啾脉冲首先展宽到一个极大值，主要由于初始负啁啾的作用，尽管色散效应加强了该作用；接着，脉冲压缩到一个极小值，伴随着色散波从主脉冲分离出去，主要由于非线性效应的作用；对应给定的光功率，脉冲展宽和压缩的次数越多，脉冲波形越接近双曲正割脉冲波形。

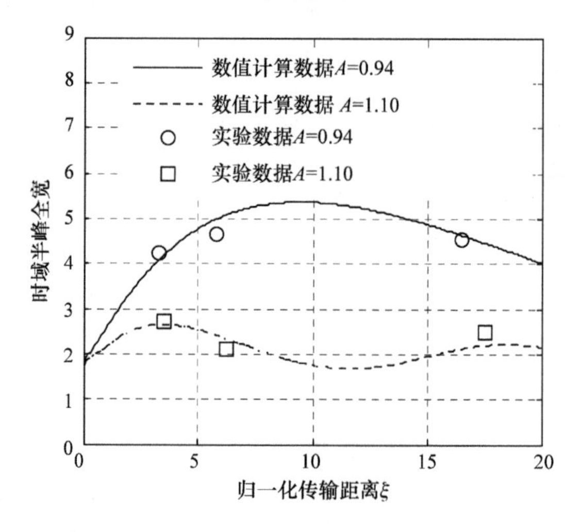

图 4-13　脉冲时域半峰全宽随传输距离的变化

图 4-14 所示是 A=0.94 时脉冲在 355m(a)和 1km(b)处的时域波形，图 4-15 所示是 A=1.10 时脉冲在 355m(a)和 1km(b)处的时域波形。两图与图 4-9(a)坐标、图例相同。由图得到，非线性传输过程中，啁啾孤子时域波形在双曲正割脉冲波形附近变化，图 4-14(a)、(b)和图 4-15(a)中时域波形为脉冲压缩到极小值之前色散波尚未从主脉冲分离出去的情况，图 4-15(b)中时域波形为脉冲压缩到极小值之后色散波已经从主脉冲分离出去一次的情况。

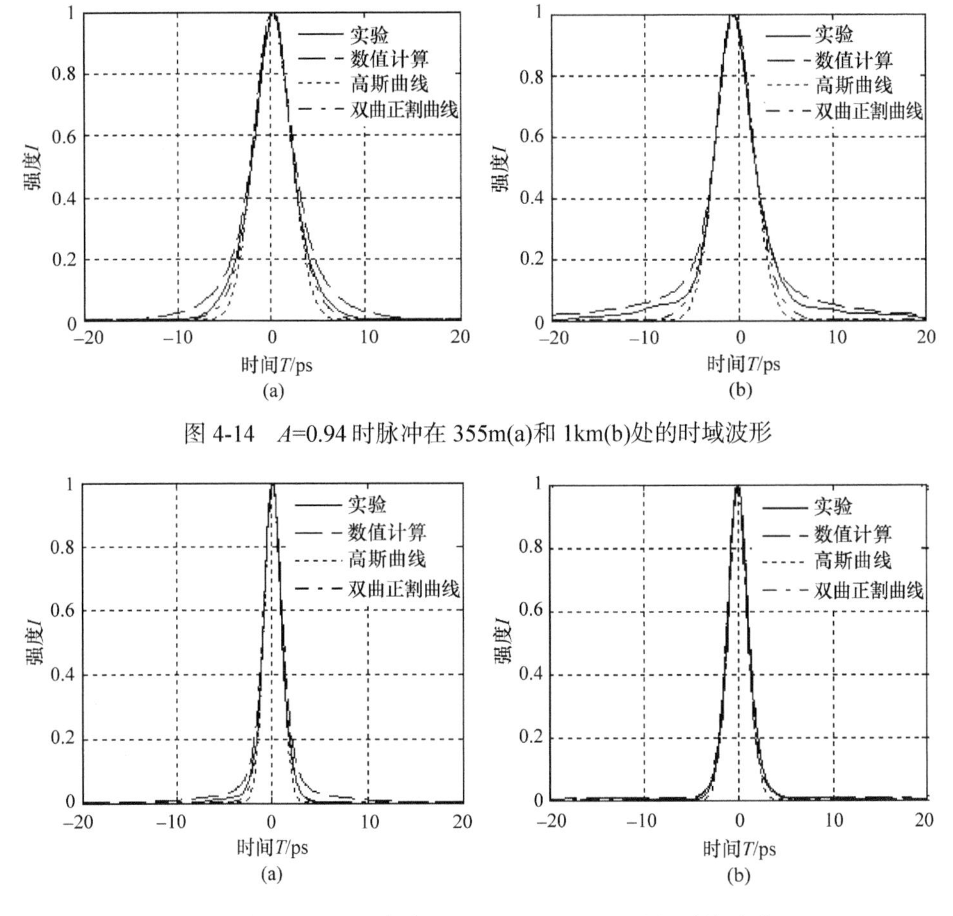

图 4-14　A=0.94 时脉冲在 355m(a)和 1km(b)处的时域波形

图 4-15　A=1.10 时脉冲在 355m(a)和 1km(b)处的时域波形

4.3　啁啾指数脉冲的非线性传输特性

下面数值研究初始线性啁啾双边指数脉冲在单模光纤反常色散区的非线性传输特性，并与双曲正割脉冲的相应特性作比较。

4.3.1　指数脉冲非线性传输的数学模型

指数脉冲在单模光纤反常色散区的非线性传输满足归一化薛定谔方程(4-4)。作者采用分步傅里叶方法，按照方程(4-4)对指数脉冲的非线性传输特性进行了数值研究，并与双曲正割脉冲的非线性传输特性作了比较，输入指数和双曲正割脉冲分别为

$$u(0,\tau)=A\exp(-|\tau|)\exp(-0.5\mathrm{i}C\tau^2),\quad u(0,\tau)=A\,\mathrm{sech}(\tau)\exp(-0.5\mathrm{i}C\tau^2) \tag{4-5}$$

式中，A 为电场 u 的振幅，$\tau=T/T_0$ 是归一化时间。在下面的数值研究中，4.3.2 节令 A=1，4.3.3 节令 A=2。

4.3.2　A=1 时指数脉冲的非线性传输特性

4.3.2.1　脉冲时间带宽积的演化

图 4-16 所示为 A=1 时脉冲时域半峰全宽(FWHM)随传输距离 ξ 的变化。曲线 1～3 分别对应指数脉冲 $C=0$、1 和−1 的情况，曲线 4～6 分别对应双曲正割脉冲 $C=0$、1 和−1 的情况。横坐标为传输距离 ξ，归一化到色散长度；纵坐标为时域 FWHM，归一到脉冲半宽度 T_0。由图 4-16 可见，在 $-2\leqslant C\leqslant 0$ 时，指数脉冲时域 FWHM 随 ξ 的增加而单调展宽，随啁啾参量 $|C|$ 的增加而增大；在 $0<C\leqslant 2$ 时，脉冲时域 FWHM 首先经历一个初始压缩阶段，然后随 ξ 和啁啾参量 C 的增加而增大；啁啾参量 C 越大，在 $\xi=1$ 附近的初始压缩越强烈。这表明初始正啁啾能诱导指数脉冲压缩。

指数脉冲时域宽度随传输距离增大和啁啾参量 $|C|$ 增大的展宽速度比双曲正割脉冲的要大，负啁啾对应的脉冲展宽速度比正啁啾的要快。这表明指数脉冲受啁啾的影响比双曲正割脉冲更敏感，负啁啾对脉冲展宽的影响比正啁啾大。指数脉冲非线性传输时的展宽比线性传输时要慢。因为脉冲压缩对应峰值的增加，所以脉冲峰值 $|u|_{\max}$ 随传输距离的变化与时域 FWHM 的变化相反。

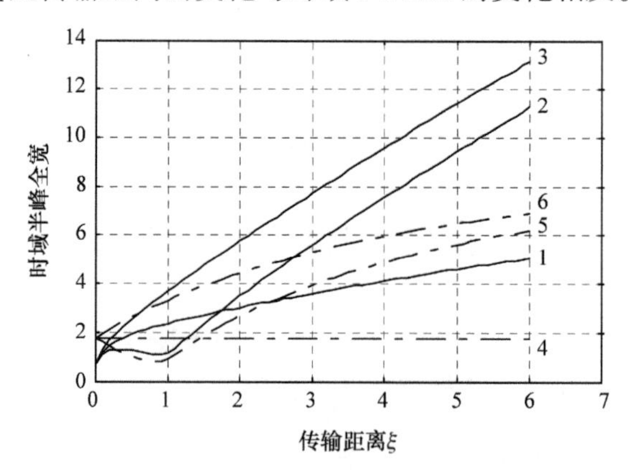

图 4-16　A=1 时光脉冲时域 FWHM 随归一化传输距离的变化

实线 1～3 所示对应指数脉冲 $C=0$、1 和−1，点划线 4～6 所示对应双曲正割脉冲 $C=0$、1 和−1

4.3.2.2　频谱半峰全宽随传输距离的变化

图 4-17 所示为脉冲频谱 FWHM 随传输距离 ξ 的变化。曲线 1～3 分别对应

指数脉冲 $C=0$、1 和−1 的情况，曲线 4～6 分别对应双曲正割脉冲 $C=0$、1 和−1 的情况。纵坐标为频谱 FWHM，归一化到 $1/T_0$。数值结果表明，在脉冲展宽过程中，在初始频率啁啾、色散和非线性的共同作用下，指数脉冲的频谱形状发生改变，但仍呈单峰形状，其边缘具有较小的振荡结构。在 $-2\leqslant C\leqslant 0$ 时，频谱 FWHM 随传输距离和啁啾参量$|C|$的增加而单调减小；在 $0<C\leqslant 2$ 时，先经历一个初始展宽阶段，然后随传输距离的增加而减小，随啁啾参量 C 的增加而增加。啁啾参量 C 值越大，频谱 FWHM 在 $\xi=1$ 附近的初始展宽越宽。频谱初始展宽和脉冲初始压缩主要由初始正啁啾诱导产生，尽管非线性增强了该效应。对于相同的初始啁啾$|C|$值，正啁啾对应的频谱 FWHM 比负啁啾的宽；正啁啾诱导的频谱展宽以后，正负啁啾对应的频谱宽度都随传输距离的增加而逐渐减小。而频谱初始压缩和脉冲初始展宽主要由初始负啁啾诱导产生，非线性减弱了该效应。

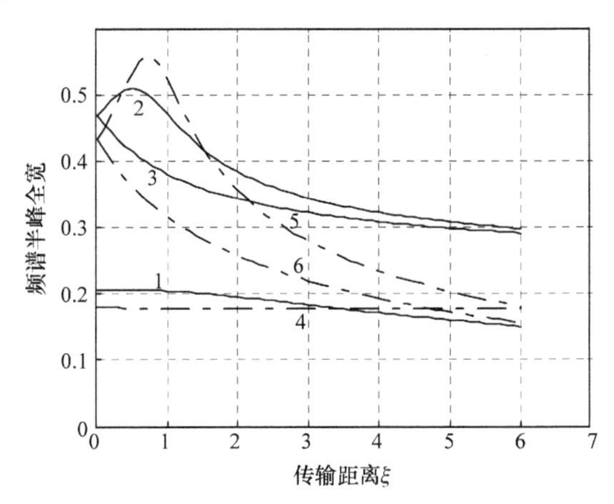

图 4-17　A=1 时光脉冲频谱 FWHM 随传输距离的变化

实线 1～3 所示对应指数脉冲 $C=0$、1 和−1，点划线 4～6 所示对应双曲正割脉冲 $C=0$、1 和−1

4.3.2.3　指数脉冲时域波形的演化

图 4-18 所示为 C=0 和 $\xi=3$ 时指数脉冲的时域波形。实线所示为指数脉冲非线性传输时的数值波形，虚线所示为与数值波形具有相同的时域 FWHM 的指数脉冲波形，点线所示为相应的高斯脉冲波形，点划线所示为相应的双曲正割脉冲波形。横坐标是归一化时间，纵坐标是时域波形的归一化强度。

数值结果表明，无啁啾指数脉冲在 $\xi=1$ 时演化为具有色散波的近高斯脉冲，在 $\xi=3$ 时演化为近双曲正割脉冲，如图 4-18 实线所示，在随后传输过程中最终演化成近双曲正割脉冲波形。

当 $0<|C|\leqslant 0.5$ 时，指数脉冲在 $\xi=1$ 时演化为具有色散波的近高斯脉冲，在

$\xi=5$ 时演化为具有色散波的近双曲正割脉冲；当 $0.5<|C|\leqslant 1$ 时，指数脉冲在 $\xi=6$ 时演化为具有色散波的近双曲正割脉冲，在随后传输过程中最终演化成近双曲正割脉冲波形；当 $1<|C|\leqslant 2$ 时，指数脉冲在传输距离更长时演化为具有色散波的近双曲正割脉冲。啁啾参量$|C|$值越大，指数脉冲演化为近双曲正割脉冲所需的传输距离越长，波形边缘的色散波越大。

无啁啾双曲正割脉冲是非线性传输方程(4-4)的解析解，是一个稳定的传输模式，这与数值结果一致。然而，指数脉冲不是非线性传输方程的解析解，不能稳定传输，其最后将演化成近双曲正割脉冲；但是在短距离范围内，$A\leqslant 1.2$ 的情况不能演化成光孤子，因为除了在 $\xi=1$ 附近正啁啾诱导的初始压缩外，脉冲随传输距离一直展宽。

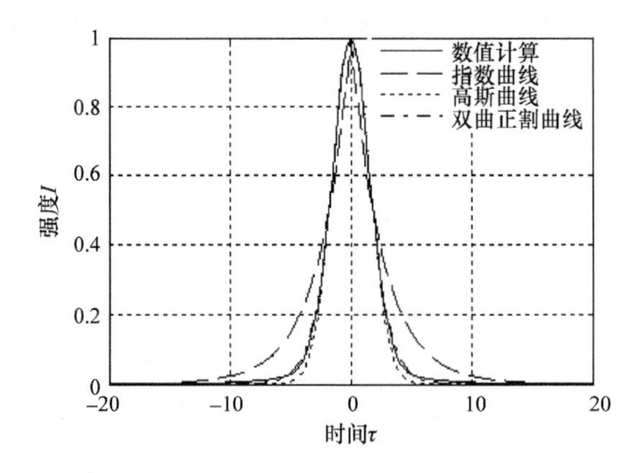

图 4-18　当 C=0 和 $\xi=3$ 时线性啁啾指数脉冲的时域波形

实线所示是指数脉冲非线性传输时的数值波形，虚线所示是与数值波形具有相同的时域 FWHM 的指数脉冲波形，点线所示是相应的高斯脉冲波形，点划线所示是相应的双曲正割脉冲波形

4.3.3　A=2 时指数脉冲的非线性传输特性

4.3.3.1　脉冲时域宽度随传输距离的变化

数值结果表明，当 A>1.2 时，指数脉冲时域 FWHM 随传输距离的增加出现了较明显的衰减振荡。图 4-19 所示为 A=2 时脉冲时域 FWHM 随归一化传输距离的变化。纵坐标是时间半峰全宽，归一化到 T_0，曲线 1～5 分别对应指数脉冲 $C=0$、-1、-2、1 和 2 的情况。由图 4-19 可见，当$|C|\leqslant 2$时，脉冲时域 FWHM 随传输距离增加出现了衰减振荡，振荡周期和振幅随啁啾参量$|C|$的增加而增大；对于相同的初始啁啾$|C|$值，正负啁啾对应的时域 FWHM 以相同的方式衰减振荡，仅仅在 $\xi=1$ 附近相差一个初始延迟。A 数值越大，振荡周期越小。脉冲展宽明显小于 A=1 的情况。脉冲峰值$|u|_{\max}$的变化与时域 FWHM 的变化相反。

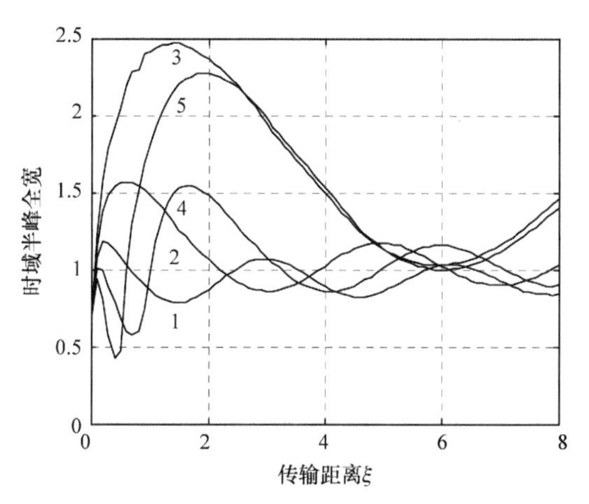

图 4-19　A=2 时光脉冲时域 FWHM 随归一化传输距离的变化

实线 1～5 所示对应指数脉冲 C = 0、−1、−2、1 和 2

4.3.3.2　指数脉冲频谱的演化

图 4-20 所示为指数脉冲在 A=2 和 C = −1 时，频谱随传输距离的演化。x 坐标是归一化传输距离，y 坐标是归一化频率，z 坐标是脉冲频谱强度，任意单位。

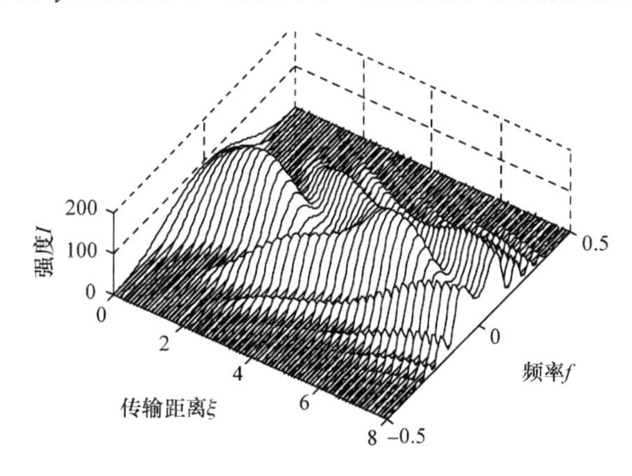

图 4-20　当 A=2 和 C = −1 时指数脉冲频谱随传输距离的演化

由图 4-20 可见，指数脉冲频谱形状在非线性传输时周期性变化，在 $\xi \leqslant 2$ 时，演化成具有振荡结构的单峰形状；在 $2 < \xi < 4$ 时，劈裂成具有更大振荡结构的双峰形状；在 $4 \leqslant \xi < 6$ 时，又演化成具有振荡结构的单峰形状；在 $6 \leqslant \xi \leqslant 8$ 时，再次劈裂成具有更大振荡结构的双峰形状；脉冲频谱变化周期随$|C|$的增加而增大，这与脉冲宽度的变化相联系。当脉冲展宽到一个极大值附近时，指数脉

冲频谱形状演化成具有振荡结构的单峰形状；当脉冲压缩到一个极小值附近时，指数脉冲频谱形状劈裂成具有振荡结构的双峰形状。正啁啾时脉冲初始压缩以后，$|C|\leqslant 2$ 时指数脉冲频谱的演化规律与 $C=-1$ 时类似。

4.3.3.3 指数脉冲波形的演化

图 4-21 所示为指数脉冲在 A=2 和 $C=-1$ 时时域波形随传输距离的演化。x 坐标是归一化传输距离，y 坐标是归一化时间，z 坐标是脉冲归一化强度。由图 4-21 可见，指数脉冲首先在 $\xi=0.6$ 处展宽到一个极大值，这主要由于初始负啁啾诱导，尽管色散增强了该效应，此时时域波形已演化成近双曲正割脉冲，脉冲边缘下降比双曲正割曲线稍慢，并具有较小的色散波，如图 4-22(a)所示。然后，脉冲在 $\xi=3$ 处压缩到一个极小值，这主要由于非线性效应诱导，色散减弱了该效应，此时时域波形演化成近双曲正割脉冲，脉冲边缘下降比双曲正割曲线稍快，较小的色散波从主脉冲中分离出去，如图 4-22(b)所示。在两个极值点的中点，脉冲波形最接近双曲正割曲线，如图 4-22(c)所示。接着，脉冲再次展宽和压缩，脉冲边缘的色散波再次从主脉冲中分离出去；脉冲展宽和压缩的次数越多，色散波越小，脉冲时域波形越接近双曲正割曲线。最后，脉冲演化成光孤子。除了初始正啁啾诱导脉冲初始压缩外，$|C|\leqslant 2$ 时指数脉冲时域波形演化规律与 $C=-1$ 时的类似。A 的数值越大，时域波形振荡周期越小，脉冲演化成光孤子所需的传输距离越小。

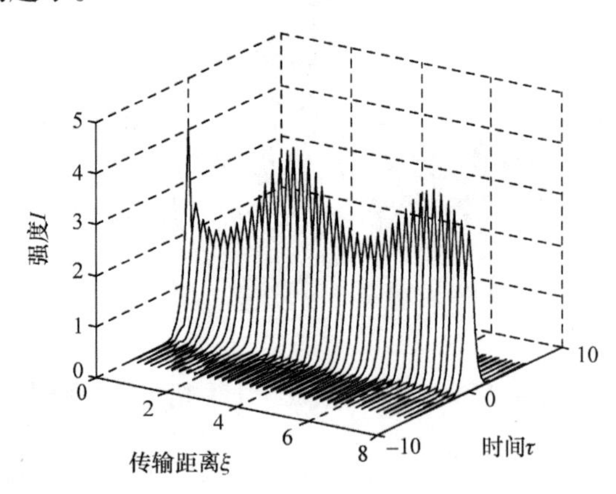

图 4-21 当 A=2 和 $C=-1$ 时指数脉冲时域波形随传输距离的演化

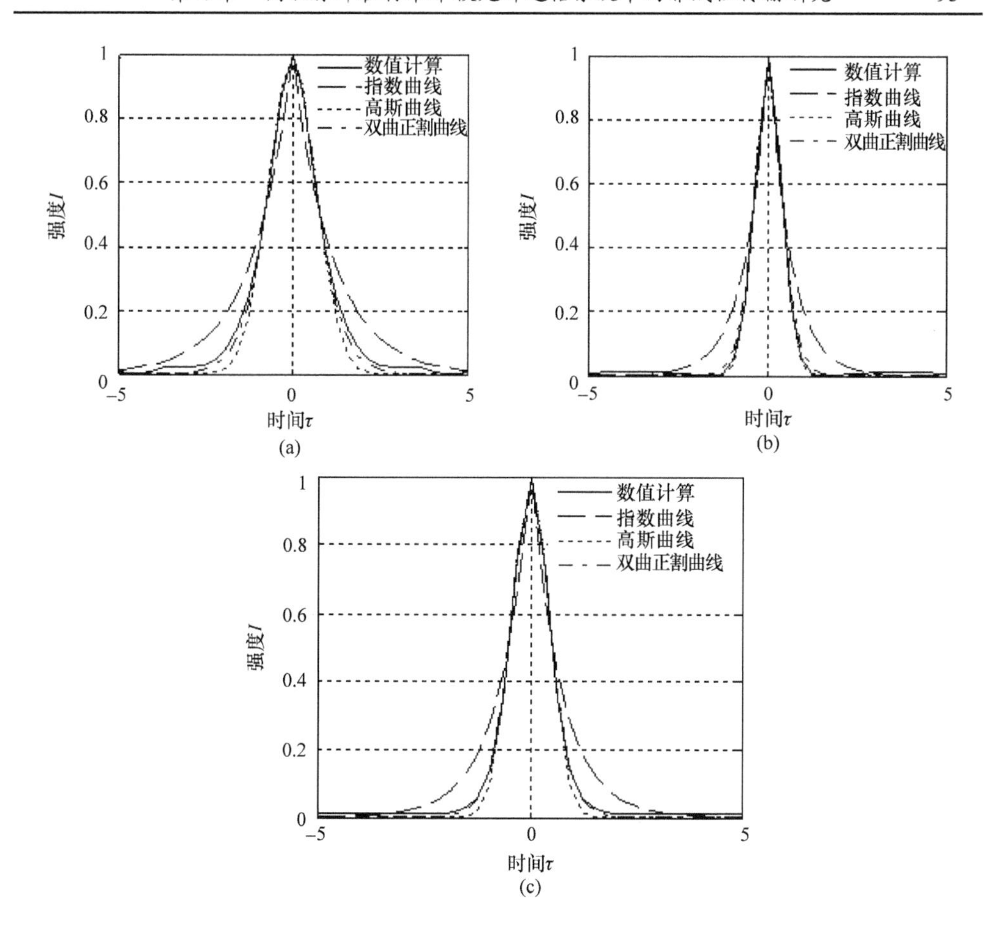

图 4-22　当 A=2 和 C= −1 时线性啁啾指数脉冲传输到 $\xi = 0.6$ (a)、$\xi = 3$ (b)、$\xi = 4$ (c)的时域波形

4.4　本章小结

利用能够准确测量脉冲时域波形等特性的 SHG-FROG 技术从实验上研究了啁啾脉冲演化形成孤子的规律和特点[36]，实验结果表明：输入脉冲传输大约 3 个色散长度时，其时域宽度、啁啾和时间带宽积都随着输入功率的增加而减小。当输入功率大于一阶孤子功率的理论值时，输入脉冲在经过大约 3 个色散长度距离后的传输过程中其宽度基本保持不变，能够演化形成孤子，其脉宽随输入功率的增加而缓慢减小。当输入功率小于一阶孤子功率的理论值时，不能演化形成孤子。

采用分步傅里叶方法数值研究了啁啾孤子脉冲的形成和传输[37]。研究表明，数值结果与实验数据一致。脉冲经过大约 3 个色散长度能够演化成啁啾孤子，脉冲时域半峰全宽和啁啾随功率的增加而逐渐减小。啁啾孤子在传输过程中，脉冲时域波形在双曲正割脉冲波形附近变化，其时域半峰全宽随传输距离的增加产生衰减振荡，其振幅和周期随功率(A)的增加而减小。

采用分步傅里叶方法数值研究了初始线性啁啾指数脉冲的非线性传输特性[38]。在 A=1 的情况下，指数脉冲受啁啾的影响比双曲正割脉冲更敏感，负啁啾对脉冲时域宽度展宽的影响比正啁啾大；脉冲展宽过程中，频谱随传输距离增加而变化，其形状为在边缘具有振荡结构的单峰。A=2 的指数脉冲在演化过程中呈明显的周期性衰减振荡，振荡周期和振幅随啁啾$|C|(|C|\leqslant 2)$的增大而增大，最后能演化成光孤子。A 越大，$|C|$越小，指数脉冲演化成光孤子所需的距离越短。相同传输距离情况下，指数脉冲演化成光孤子所需的最小振幅 A 比啁啾双曲正割脉冲的大。指数脉冲频谱形状呈周期性变化。本章的研究为孤子源及其通信系统的设计优化提供了重要依据。

参考文献

[1] 阿戈沃. 非线性光纤光学原理及应用. 贾东方, 余震虹, 谈斌, 译. 北京: 电子工业出版社, 2002.

[2] 刘颂豪, 赫光生. 强光光学及其应用. 广州: 广东科技出版社, 1995.

[3] 庞小峰. 孤子物理学. 北京: 科学出版社, 1987.

[4] 杨祥林, 温扬敬. 光纤孤子通信理论基础. 北京: 国防工业出版社, 2000.

[5] 黄景宁, 徐济仲, 熊吟涛. 孤子概念、原理和应用. 北京: 高等教育出版社, 2004.

[6] Mollenauer L F, Stolen R H, Gordon J P. Experimental observation of picosecond pulse narrowing and solitons in optical fiber. Phys. Rev. Lett., 1980, 45(13): 1095～1098.

[7] Gouveia-Neto A S, Wigley P G J, Taylor J R. Soliton generation through Raman amplification of noise bursts. Optics Letters, 1989, 14(20): 1122～1124.

[8] Iwatsuki K, Suzuki K I, Nishi S. Adiabatic soliton compression of gain-switched DFB-LD pulse by distributed fiber Raman amplification. IEEE Transactions Photonics Technology Letters, 1991, 3(12): 1074～1076.

[9] Murphy T E. 10-GHz 1.3-ps pulse generation using chirped soliton compression in a Raman gain medium. IEEE Photonics Technology Letters, 2002, 14(10): 1424～1426.

[10] Mollenauer L F, Stolen R H, Islam M N. Experimental demonstration of soliton propagation in long fibers: Loss compensated by Raman gain. Optics Letters, 1985, 10(5): 229～231.

[11] Iwatsuki K, Nishi S, Saruwatari M, et al. 5Gb/s optical soliton transmission experiment using Raman amplification for fiber-loss compensation. IEEE Photonics Technology Letters, 1990, 2(7): 507～509.

[12] Okhrimchuk A G, Onishchukov G, Lederer F. Long-haul soliton transmission at 1.3 μm using distributed Raman amplification. Journal of Lightwave Technology, 2001, 19(6): 837～841.

[13] 高以智, 姚敏玉, 许宝西, 等. 2.5GHz 光孤子传输. 高技术通讯, 1994, 7: 4～6.

[14] 许宝西, 李京辉, 姜新, 等. 2.5GHz 光孤子传输. 电子学报, 1995, 23(11): 38～54.

[15] 杨祥林, 毛庆和, 温扬敬, 等. 30km 2.5GHz 光孤子波传输与压缩实验研究. 高技术通讯, 1996,10: 26～28.

[16] 余建军, 杨伯君, 余建国, 等. 光孤子传输实验研究. 光电子 · 激光, 1996, 7(5): 267～272.

[17] 余建军, 杨伯君, 管克俭. 5GHz 的 16.2ps 超短光脉冲的产生. 光学学报, 1998, 18(1) :14～17.

[18] 余建军, 杨伯君, 管克俭. 基于不同色散光纤的光纤链的孤子传输研究. 光学学报, 1998, 18(4): 446～450.

[19] 张晓光, 林宁, 张涛, 等. 预啁啾 10GHz, 38km 色散管理孤子的传输实验. 光子学报, 2001, 30(7): 813～817.

[20] Wang S, Wang Y B, Feng G Y, et al. Generation of double-scale pulses in a LD-pumped Yb:phosphate solid-state laser. Appl. Opt., 2017, 56: 897～900.

[21] Marec A L, Guilbaud O, Larroche O, et al. Evidence of partial temporal coherence effects in the linear autocorrelation of extreme ultraviolet laser pulses. Opt. Lett., 2016, 41: 3387～3390.

[22] Chaparro A, Furfaro L, Balle S. Subpicosecond pulses in a self-starting mode-locked semiconductor-based figure-of-eight fiber laser. Photon. Res., 2017, 5: 37～40.

[23] Lauterio-Cruz P, Hernandez-Garcia J C, Pottiez O, et al. High energy noise-like pulsing in a double-clad Er/Yb figure-of-eight fiber laser. Opt. Express, 2016, 24: 13778～13787.

[24] Lin J H, Chen C L, Chan C W, et al. Investigation of noise-like pulses from a net normal Yb-doped fiber laser based on a nonlinear polarization rotation mechanism. Opt. Lett., 2016, 41: 5310～5313.

[25] Zhang F, Fan X W, Liu J, et al. Dual-wavelength mode-locked operation on a novel Nd^{3+}, Gd^{3+}: SrF_2 crystal laser. Opt. Mater. Express, 2016, 6: 1513～1519.

[26] Chao M S, Cheng H N, Fong B J, et al. High-sensitivity ultrashort mid-infrared pulse characterization by modified interferometric field autocorrelation. Opt. Lett., 2015, 40: 902～905.

[27] Sun B S, Salter P S, Booth M J. Pulse front adaptive optics: A new method for control of ultrashort laser pulses. Opt. Express, 2015, 23: 19348～19357.

[28] Lin S S, Hwang S K, Liu J M. High-power noise-like pulse generation using a 1.56-μm all-fiber laser system. Opt. Express, 2015, 23: 18256～18268.

[29] Traore A, Lalanne E, Johnson A M. Determination of the nonlinear refractive index of multimode silica fiber with a dual-line ultra-short pulse laser source by using the induced grating autocorrelation technique. Opt. Express, 2015, 23: 17127～17137.

[30] Suzuki M, Ganeev R A, Yoneya S, et al. Generation of broadband noise-like pulse from Yb-doped fiber laser ring cavity. Opt. Lett., 2015, 40: 804～807.

[31] Tian W L, Wang Z H, Liu J X, et al. Dissipative soliton and synchronously dual-wavelength mode-locking Yb:YSO lasers. Opt. Express, 2015, 23: 8731～8739.

[32] MacLeod A M, Yan X, Gillespie W A, et al. Formation of low time-bandwidth product, single-sided exponential optical pulses in free-electron laser oscillators. Physical Review E, 2000, 62(3): 4216～4220.

[33] Shapiro S L. 超短光脉冲-皮秒技术及其应用. 朱世清, 译. 北京: 科学出版社, 1987.

[34] 朱京平. 光电子技术基础. 成都: 四川科学技术出版社, 2003.
[35] Satsuma J, Yajima N. Initial value programs of one-dimensional self-modulation of nonlinear waves in dispersive media. Prog. Theor. Phys. Suppl., 1974, 55: 284～306.
[36] 刘山亮, 郑宏军. 光脉冲在标准单模光纤中演化形成孤子的实验研究. 光学学报, 2006, 26(9): 1313～1318.
[37] Zheng H J, Liu S L. Formation and propagation of chirped optical soliton. Proc. of SPIE, 2006, 6353: 635323-1-5.
[38] 郑宏军, 刘山亮, 徐静平. 啁指数啾脉冲的非线性传输研究. 光通信研究, 2007, 140(2): 11～13.

第5章　啁啾脉冲在特种光纤通信系统中传输特性的研究

在光纤反常色散区发现光纤孤子以来[1]，光脉冲的传输得到了广泛的研究，但是其研究工作大多限于反常色散区[2-7]。近年来，光脉冲在正常色散区的传输研究受到关注，相关文献数值计算了光脉冲在正常色散区的传输，在单模光纤正常色散区研究具有潜在应用价值的暗孤子[6]，用正常色散光纤(即负色散光纤，色散补偿光纤(DCF))消除光脉冲啁啾和光脉冲压缩[7-9]，在正常色散平坦光纤中形成超连续谱[10, 11]，利用正常色散光纤产生飞秒激光[12]。上述在正常色散区的实验研究大多采用自相关技术测量脉冲时域，不能测得光脉冲的波形等参量，其研究受到限制。然而，二次谐波频率分辨光学门技术能够准确测量光脉冲时域波形、相位、啁啾等参量[13-18]，如 Barry 研究组采用二次谐波频率分辨光学门技术测量了光功率脉冲经短距离(20m 和 100m)色散位移光纤后的特性，主要演示了二次谐波频率分辨光学门技术恢复输出光脉冲强度、相位等参量信息的实验[18]。又考虑到采用二次谐波频率分辨光学门技术研究光脉冲在色散平坦光纤正常色散区传输罕有报道，本章采用能够准确测量光脉冲波形的二次谐波频率分辨光学门技术，对光脉冲在正常色散的色散平坦光纤中的传输进行了实验研究，得到了光脉冲在不同输入功率下，经不同长度、不同色散值的正常色散平坦光纤中传输前后脉冲宽度、波形、光谱等的变化规律和特点，这对进一步研究和利用光脉冲在正常色散区的特性是非常有益的探索。

5.1　啁啾脉冲在色散平坦光纤正常色散区的传输特性

5.1.1　实验装置

实验装置如图 5-1 所示，分布反馈式激光器(DFB laser)是平均输出光功率为 6dBm 和中心波长为 1551.9nm 的激光器；40GHz 高速电光调制器由德国 U2T 公司生产；81250 20GHz CLK 是安捷伦公司的高速误码仪产生的高速电时钟；40GHz 高速电放大器由德国 U2T 公司生产；EDFA 为法国 KPS 公司的掺铒光纤放大器，光放大波段为 1535～1565nm。SHG-FROG 为新西兰南方光子有限公司的 SHG-FROG 脉冲仪，能够测量得到光脉冲时域宽度、谱宽、波形、相位和啁啾等各种特征信息；AQ6319 是日本横河公司的高精度光谱分析仪。在本章所述实验中 DFB 激光器输出的激光光波经偏振控制器进入高速电光调制器，被放大后的 21.6GHz 电时钟调制成高速光脉冲，光脉冲经掺铒光纤放大器放大后输入具有不同色散的光纤进行传输。采用 SHG-FROG 和 AQ6319 对在光纤中传输前后的光脉冲进行测量和分析。各器件之间用光纤活动连接器或跳线连接。实验中所用两条正常色散光纤为长飞光纤光缆有限公司生产的线性色散平坦光纤，其中一条光纤模场直径为 7.3μm、1550nm 处色散参量 $D = -0.187$ps/(nm · km)，色散斜率为 0.01587834ps/(nm^2 · km)，光纤损耗为 0.227dB/km，长度是 4.265km；另一条光纤模场直径为 7.41μm、色散参量 $D = -0.746$ps/(nm · km)，色散斜率为 0.006480361ps/(nm^2 · km)，光纤损耗为 0.229dB/km，长度是 6.346km。

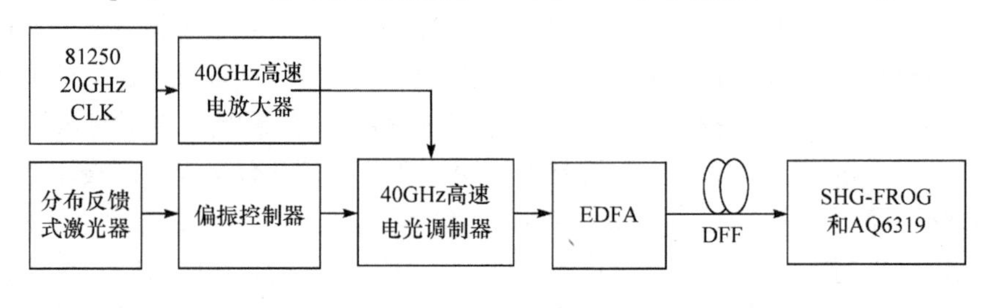

图 5-1　实验装置

5.1.2　实验结果与分析

首先采用 SHG-FROG 脉冲分析仪测量得到输入光脉冲参量(EDFA 泵浦电流 2A 时(光功率 389mW)的光脉冲，未在光纤中传输)。图 5-2(a)所示是测量得到的脉冲的时域波形曲线。图 5-2 中横坐标是时间 T，单位 ps，纵坐标是归一化的脉冲强度。图 5-2(a)中实点是实验测量得到的脉冲波形曲线，虚线是高斯脉冲曲线，点划线是双曲正割脉冲曲线，实线是余弦型脉冲波形曲线。由图 5-2(a)可见，输入脉冲波形与余弦脉冲曲线吻合得较好，脉冲前后沿比双曲正割和高斯脉冲曲线下降得都快。测量结果表明输入光脉冲的半峰全宽为 23.01ps，随着 EDFA 泵浦功率的增加(光功率 646mW)，输入脉冲的时域宽度几乎不变。图 5-2(b)所示是输入光脉冲的相位曲线，横坐标是时间 T，纵坐标表示脉冲相位，单位 rad。实点是实验测量得到的脉冲相位曲线，实线所示是线性啁啾参量 $C = -1.5$ 时的理论相位曲线。由图 5-2(b)可见，实验脉冲呈线性啁啾。

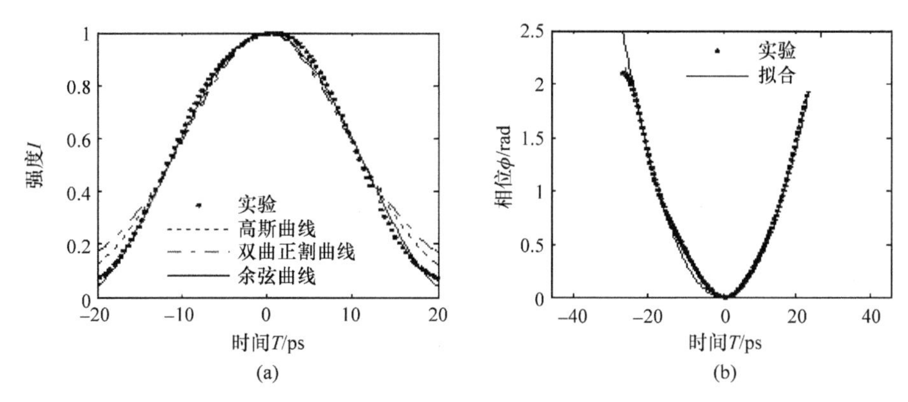

图 5-2　传输前测量得到的光脉冲时域波形曲线(a)和相位曲线(b)

(a)实点是实验测量得到的脉冲波形曲线，虚线是高斯脉冲曲线，点划线是双曲正割脉冲曲线，实线是余弦型脉冲波形曲线；(b)实点是实验测量得到的脉冲相位曲线，实线是线性啁啾参量 $C=-1.5$ 时的理论相位曲线

5.1.2.1　光脉冲时域宽度与输入功率的关系

在图 5-1 所示实验装置中，通过调节 EDFA 泵浦激光器的工作电流改变输入光纤脉冲的功率，EDFA 输出的不同功率的光脉冲经光纤传输后输入 HR200 进行测量和分析。图 5-3 所示是接入不同色散、不同长度的正常色散光纤时，实验测量得到的输出光脉冲时域半峰全宽随输入平均光功率的变化关系。图 5-3 中横坐标是 EDFA 输出光功率(两个泵浦电流相等，工作电流的增加对应光功率的增大)，单位 mW；纵坐标是光脉冲时域半峰全宽，单位 ps。从图 5-3 中可以看出，初始负啁啾光脉冲在色散平坦光纤正常色散区传输后，输出脉冲时域宽度均比输入脉冲宽度要小，与初始负啁啾光脉冲在反常色散区中初始展宽不同；对于确定的光纤长度，输出脉冲的时域宽度随着输入光脉冲平均功率的增加而减小。

这表明，与色散引起的脉冲展宽相比，本实验中非线性引起的脉冲压缩占主导地位。当输入功率较小，色散的作用大于非线性的作用，即 $L_D=\dfrac{T_0^2}{|\beta_0''|}<L_{\mathrm{NL}}=1/\gamma P_0$ 时，脉冲压缩的原因主要是负啁啾的光脉冲与光纤色散的乘积 $C\beta_0''<0$，经过一段光纤后达到变换极限，脉冲压缩到最窄(当大于变换极限长度后，继续展宽)；反之，当 $L_D=\dfrac{T_0^2}{|\beta_0''|}\gg L_{\mathrm{NL}}=1/\gamma P_0$ 时，非线性起主要作用。在 $L_D=\dfrac{T_0^2}{|\beta_0''|}$ 略大于 $L_{\mathrm{NL}}=1/\gamma P_0$ 时，二者共同起作用，它们之比 $N^2=L_D/L_{\mathrm{NL}}=1$ 时，会形成暗孤子。当 $N^2=L_D/L_{\mathrm{NL}}>2$ 时，会形成高阶暗孤子，从而经历一个脉冲压缩期(但也会经历脉冲展宽期)。

在相同输入功率下，经过不同色散光纤后的输出脉冲宽度不同。色散较小光纤 DFF AA0 的输出脉冲宽度比色散较大光纤 DFF EA0 的输出脉冲宽度要窄，尽管后者比前者要长。这表明，在相同的输入功率下，光纤色散越小，非线性越占主导作用，其引起的脉冲压缩越强。

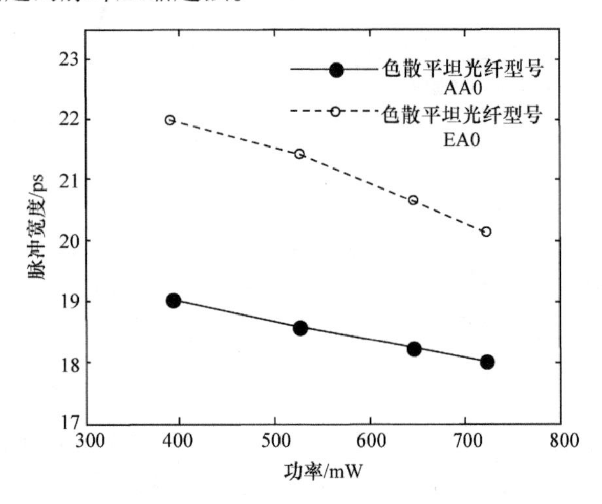

图 5-3 脉冲宽度在不同光纤 DFF AA0 和 DFF EA0 随输入功率的变化

5.1.2.2 光脉冲波形的演化

图 5-4 所示是 EDFA 泵浦电流为 2A 时(光功率 389 mW)，输入脉冲在光纤中分别经过光纤 DFF AA0 (a)和 DFF EA0 (b)传输后输出光脉冲的波形曲线。图中实点是实验测量得到的脉冲波形曲线，实线是高斯脉冲曲线，点划线是双曲正割脉冲曲线，虚线是双边指数脉冲波形曲线。由图 5-4 可见，实验测量得到经光纤 DFF EA0 传输后的输出光脉冲宽度为 21.98ps，比经光纤 DFF AA0 传输后的输出脉冲宽度 19.02ps 宽；两输出脉冲宽度比输入光脉冲宽度 23.01ps 都窄。两输出脉冲波形曲线与近高斯脉冲曲线基本一致。这表明输入脉冲在正常色散光纤 DFF AA0 和 DFF EA0 中传输后都被非线性效应压缩，但都没有演化成为具有双曲正割波形的光纤孤子脉冲。当 EDFA 泵浦电流增加到 3A 时(光功率 646mW)，两输出脉冲波形仍然呈近高斯脉冲曲线。图 5-5 所示是 EDFA 泵浦电流为 2A 时，输入脉冲在光纤中分别经过光纤 DFF AA0(a)和 DFF EA0(b)传输后输出光脉冲的波形曲线。图中实点所示是由 SHG-FROG 实验测量得到的脉冲相位曲线，实线所示分别是线性啁啾参量 $C=-1.15$(图 5-5(a))和 $C=-2$(图 5-5(b))时的理论相位曲线。由图 5-5 可见，实验脉冲仍具有线性啁啾。其中，脉冲经小色散光纤 DFF AA0 传输后，其啁啾稍有减小；经较大色散光纤 DFF EA0 传输后，其啁啾增加。

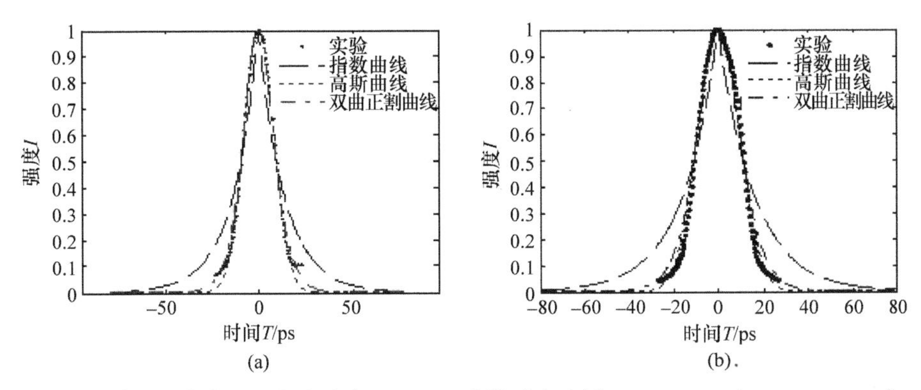

图 5-4　输入功率为 2A 时(光功率 389 mW)光脉冲在光纤 DFF AA0(a)和 DFF EA0(b)中传输后的波形曲线

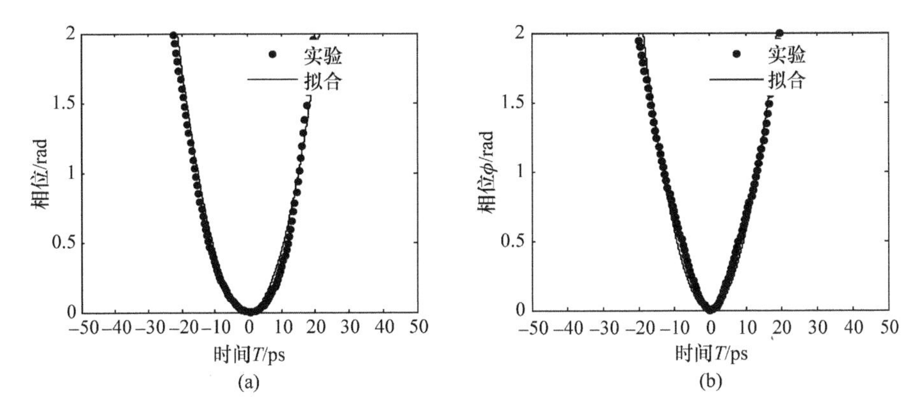

图 5-5　输入功率为 2A 时(光功率 389 mW)光脉冲在光纤 DFF AA0(a)和 DFF EA0(b)中传输后的相位曲线

5.1.2.3　脉冲光谱的演化

图 5-6 所示是输入光脉冲在光纤 DFF AA0 中传输前(a)后(b)的脉冲光谱。图中实线所示是 EDFA 泵浦电流 2A 时(光功率 389mW)的脉冲光谱曲线，虚线所示是 EDFA 泵浦电流 3A 时(光功率 646mW)的脉冲光谱曲线。由图 5-6(a)可见，实验测量得到的输入脉冲中心波长在 1551.9nm，光谱长波长区与短波长区关于中心光谱呈近似对称分布。随着 EDFA 泵浦电流由 2A 增加到 3A，光谱的各个谱峰稍微增加，但光谱没有展宽，这与输入脉冲的时域宽度 EDFA 泵浦功率增加几乎不变相关联。由图 5-6(b)可见，在光纤 DFF AA0 中传输后的脉冲光谱仍然呈近似对称分布，光谱明显展宽，在长、短波长区的 1552.9nm 和 1550.9nm 处出现谱峰。随着 EDFA 泵浦功率的增加，光谱的各个谱峰稍微增加，光谱展宽更明

显，这与在光纤中传输后的输出脉冲的时域宽度随 EDFA 泵浦功率增加逐渐减小相联系。输入光脉冲在光纤 DFF EA0 中传输后的脉冲光谱变化规律与上述结果相似。

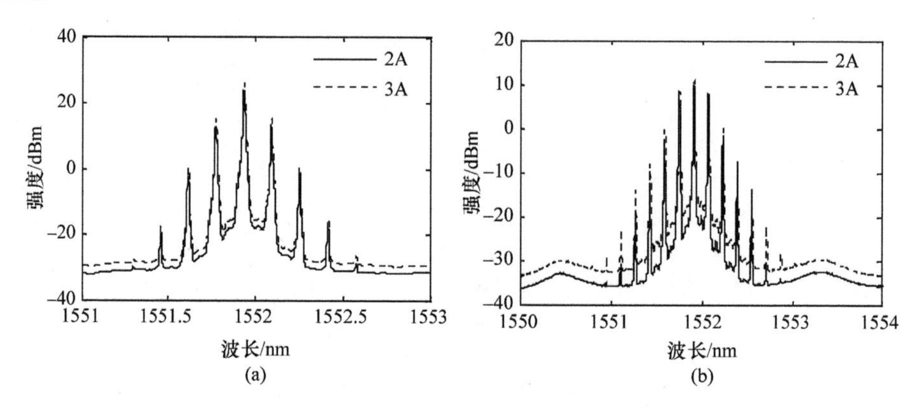

图 5-6 输入光脉冲在光纤 DFF AA0 中传输前(a)后(b)的脉冲光谱

5.1.3 讨论与结论

根据光纤中脉冲的传输理论，输入光脉冲在色散很小的光纤 DFF AA0 传输时，光脉冲传输的色散长度为

$$L_D = \frac{T_0^2}{|\beta_2|} \tag{5-1}$$

在本章所述的实验条件下，二阶色散参量 β_2=0.21ps^2/km，对半峰全宽为 23.01ps 的余弦型输入光脉冲 T_0=14.65ps，色散长度很大。在色散长度很大的情况下，脉冲在光纤中传输时，非线性效应占主导地位[6]。这与脉冲传输后脉宽变窄的实验结果相吻合。

利用二次谐波频率分辨光学门(SHG-FROG)脉冲分析仪从实验上较为细致地分析研究了具有较大啁啾光脉冲在较高输入光功率情况下、在不同长度和不同色散值的色散平坦光纤正常色散区的传输规律和特点，实验结果表明：①所研究的光纤长度输出脉冲的时域宽度随着输入光脉冲平均功率的增加而减小。②在相同输入功率下，色散较小光纤 DFF AA0 的输出脉冲宽度比色散较大光纤 DFF EA0 的输出脉冲宽度要窄。③初始负啁啾光脉冲在正常色散光纤中传输后演化成近高斯脉冲。④输出脉冲光谱仍然呈近似对称分布，输出谱宽增加并随 EDFA 泵浦功率的增加明显增加。

5.2　色散离散渐减光纤中的啁啾脉冲压缩

近年来，脉冲压缩作为一种非常重要的非线性效应，在光传输、高能强场等领域引起了越来越广泛的关注[19-33]。人们积极探索发现了很多方法可以实现脉冲压缩，如采用光纤光栅压缩脉冲[19-21]，利用高阶孤子压缩效应在标准单模光纤[22, 23]、色散位移光纤[19, 24]、色散渐减光纤[25-28]、光子晶体光纤[28-30]和梳状色散光纤等各种光纤[31-33]中进行脉冲压缩。

然而，先前的脉冲压缩实验研究通常采用不能测量脉冲时域波形、相位、啁啾等信息的脉冲自相关技术等[22,33]，这也表明要想对光脉冲压缩等相关领域进行深入研究需要采用新的技术或方法；二次谐波频率分辨光学门技术是能够准确测量脉冲时域波形、相位、啁啾等信息的新技术[14-16, 34]。为了对光脉冲压缩等相关领域进行更深入研究，本节首次采用二次谐波频率分辨光学门技术实验研究光脉冲在色散离散渐减光纤中的脉冲传输压缩特性，并与采用分步傅里叶数值计算结果作了比较。色散离散渐减光纤由多段光纤组成，虽然每一段光纤的色散值都是固定的，但是后面一段光纤的色散值总小于前面一段的色散值，即阶跃性的下降。它与色散连续渐减的光纤虽有不同，但是当每一段光纤足够短，且渐减量都相等，则也可看作连续渐减光纤，以下我们统称这类光纤为色散渐减光纤。实验中，被试光纤由三段光纤组成，一段是具有高色散值的标准单模光纤，一段是具有中等色散值的色散位移光纤，一段是具有低色散值的色散平坦光纤。首先，采用二次谐波频率分辨光学门技术实验测量输入脉冲的时域波形、相位等信息；接着，从实验数据得到输入脉冲的时域表达形式等相关参量，为下一步进行分步傅里叶数值计算研究奠定基础；最后，实验研究光脉冲在色散渐减光纤中的脉冲传输压缩特性，与采用分步傅里叶数值计算结果作比较。研究表明，实验数据与数值计算结果吻合，这是一个进行脉冲压缩的有效方案。光脉冲在色散渐减光纤中传输时能够有效压缩脉冲时域，光脉冲谱域被展宽。

5.2.1　输入光脉冲的特性实验测量

图 5-7 所示是脉冲压缩实验装置。DFB 激光器传输一个中心波长在 1551.9nm 的激光光波，该激光光波经过偏振控制器调节偏振后，进入一个 40GHz 的马赫-曾德尔调制器；安捷伦 43Gbit/s 并行误码测试仪 81250 产生 21.6GHz 时钟脉冲，该时钟脉冲经 40GHz 电放大器放大后进入马赫-曾德尔调制器；在马赫-曾德尔调制器中，时钟脉冲调制激光光波产生 21.6GHz 的激光脉冲；该激光脉冲经掺铒光纤放大器放大后作为输入脉冲进入色散渐减光纤传输压缩；可采用光谱分析仪 AQ6319

实验测量输入、输出激光脉冲光谱和利用二次谐波频率分辨光学门技术实验测量激光脉冲的时域波形、相位、脉宽等信息。

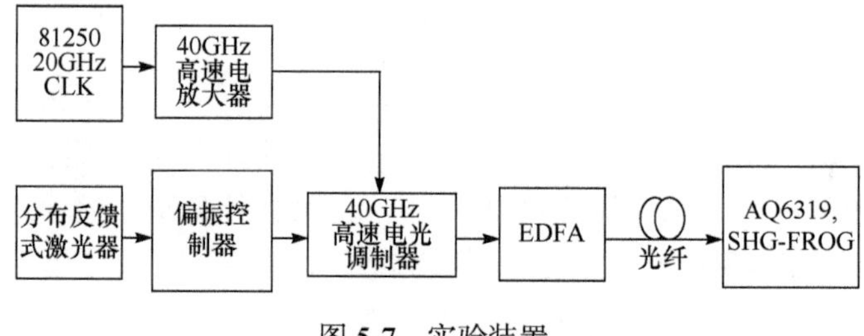

图 5-7　实验装置

首先利用二次谐波频率分辨光学门技术实验测量输入光脉冲的时域波形、相位等信息，并将实验数据导入 Matlab 程序中采用波形拟合方式获得输入光脉冲的数学表达式。图 5-8 所示是掺铒光纤放大器两个泵浦电流均为 3A(I_p=3A)时输入光脉冲的时域波形、相位信息。实点所示是实验数据，该数据与电场

$$U(0,T)=\cos\left(\frac{T}{T_0}\right)\exp\left(-\frac{\mathrm{i}CT^2}{2T_0^2}\right) \tag{5-2}$$

非常吻合。式(5-2)中，$U(0,T)$ 是输入光脉冲的瞬变电场，T 是时间，$T_0=\Delta T/1.5708=23.22/1.5708\text{ps}$ 是在 $\cos^2 1$ 强度处的半宽度，ΔT 是半峰全宽，$C=-1.5$ 是线性频率啁啾参量。图 5-9 所示是输入脉冲的光谱。实线所示是实验测量的光谱，点线是输入光脉冲式(5-2)的傅里叶变换谱。实验光谱关于中心波长基本对称，−3dB 谱宽 0.12nm。可以表明光脉冲时域波形的规则对称，实验光谱与数值计算得到的傅里叶变换谱包络一致。光谱宽度随掺铒光纤放大器光功率的增加几乎不变。

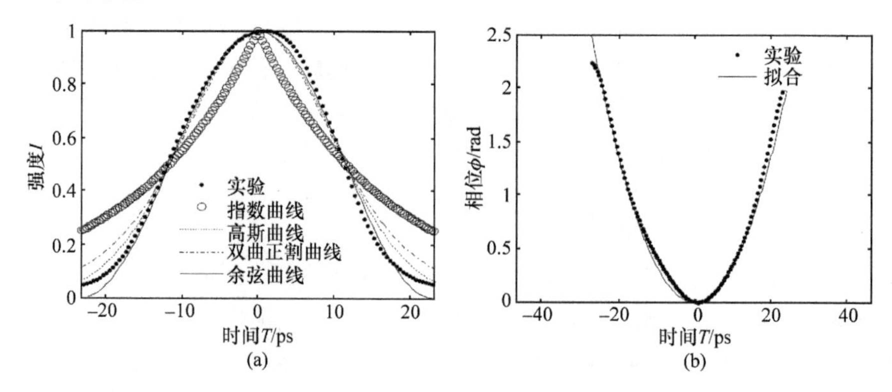

图 5-8　掺铒光纤放大器两个泵浦电流均为 3A(I_p=3A)时输入光脉冲的时域波形(a)、相位(b)信息

(b)实点所示是实验数据，实线所示是式(5-2)的相位曲线

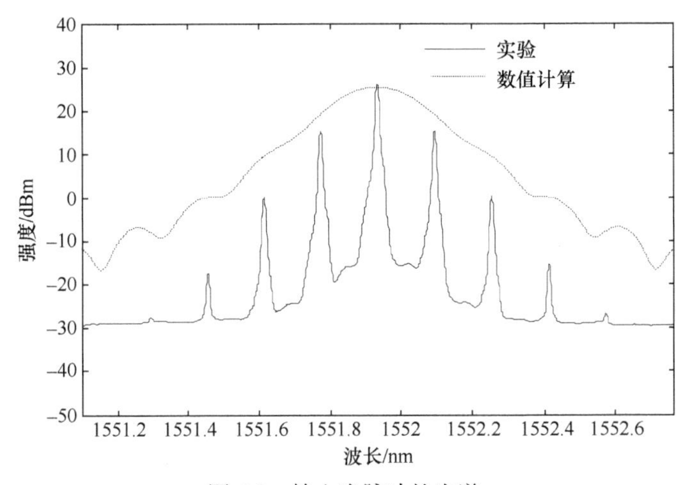

图 5-9　输入光脉冲的光谱

实线所示是实验测量的光谱，点线是输入光脉冲式(5-2)的傅里叶变换谱

5.2.2　脉冲压缩的时域特性数值计算与实验测量分析

将光脉冲输入不同的色散渐减光纤中进行脉冲压缩实验。实验和数值计算发现，并不是在所有的色散渐减光纤中都能获得好的脉冲压缩效果，当色散平坦光纤的总色散是色散位移光纤的总色散值的 1/3、色散位移光纤的总色散值是标准单模光纤总色散值的 1/3 时，可以获得良好的脉冲压缩效果。在我们的脉冲压缩实验中，标准单模光纤的模场直径为 9μm，在 1550nm 的色散参量为 14.81ps/(nm · km)，色散斜率为 0.085ps/(nm^2 · km)，光纤损耗为 0.183dB/km，光纤长度为 2.17km。色散位移光纤的模场直径为 9.5μm，在 1565nm 的色散参量为 6.02ps/(nm · km)，色散斜率为 0.077ps/(nm^2 · km)，光纤损耗为 0.199dB/km，光纤长度为 2.162km。色散平坦光纤的模场直径为 7.27μm，在 1550nm 的色散参量为 0.177ps/(nm · km)，光纤损耗为 0.226dB/km，光纤长度为 12.7km。标准单模光纤、色散位移光纤和色散平坦光纤在 1551.9nm 处的总二阶色散值分别是 -41.45ps^2、-14.26ps^2 和 -3.81ps^2。光脉冲在色散渐减光纤中脉冲压缩的理论模型可修改为 [19]

$$\mathrm{i}\frac{\partial U}{\partial \xi}+\frac{1}{2}d(\xi)\frac{\partial^2 U}{\partial \tau^2}+|U|^2 U=-0.5\mathrm{i}\alpha_s L_D U \tag{5-3}$$

式中，U 是输入光脉冲的归一化瞬变电场，ξ 是归一化传输距离，$d(\xi)=\beta_2(\xi)/\beta_2(0)$ 是色散渐减光纤的归一化色散参量，$\tau=T/T_0$ 是归一化时间，α_s 是信号波长处的光纤损耗系数，L_D 是一个色散长度。按照式(5-3)采用分

步傅里叶计算方法研究了光脉冲在色散渐减光纤中的脉冲压缩特性，数值计算的输入光脉冲是利用二次谐波频率分辨光学门技术实验测量得到的输入光脉冲式(5-2)。

图 5-10 所示是数值计算得到的输入光脉冲时域宽度随归一化传输距离(17.032km)的变化规律。实线所示是数值计算结果，实点所示是 I_p=3A 时输出光脉冲的实验数据。研究表明，具有初始负啁啾的输入光脉冲在色散渐减光纤传输压缩时，除了在刚开始传输处有一次脉冲展宽外，输入光脉冲随传输距离的增加时域逐渐被压缩。数值计算得到的光脉冲时域宽度 2.72ps 与利用二次谐波频率分辨光学门技术实验测量得到的脉冲宽度一致，都远远小于初始脉冲宽度 23.22ps。实验得到的输出脉冲时域宽度 2.42ps 的时域波形如图 5-11 所示。实点所示是光脉冲时域波形的实验数据，实线所示是数值计算得到的结果。可以得到，输出光脉冲的实验时域波形与数值计算结果一致，脉冲边缘有一点底座。输出光脉冲的实验时域宽度 2.42ps 大约是初始入射时域宽度的 1/10，是 I_p=1.5A 时输出光脉冲时域宽度 9.26ps 的 1/4。这表明啁啾光脉冲在色散渐减光纤中能够对脉冲时域有效压缩，脉冲压缩比率随输入光功率增加而增加。光脉冲时域压缩与脉冲光谱展宽相关联[19]。图 5-12 所示是 I_p=3A 时输出光脉冲的实验光谱。可以得到，I_p=3A 时输出光脉冲的−3dB 实验光谱宽度被展宽为 1.45nm，大约是 I_p=1.5A 时输出光脉冲光谱宽度 0.54nm 的 2.7 倍，是初始入射光谱宽度 0.12nm 的 12 倍多。同时表明，光脉冲的光谱展宽随输入光功率增加而迅速增加。

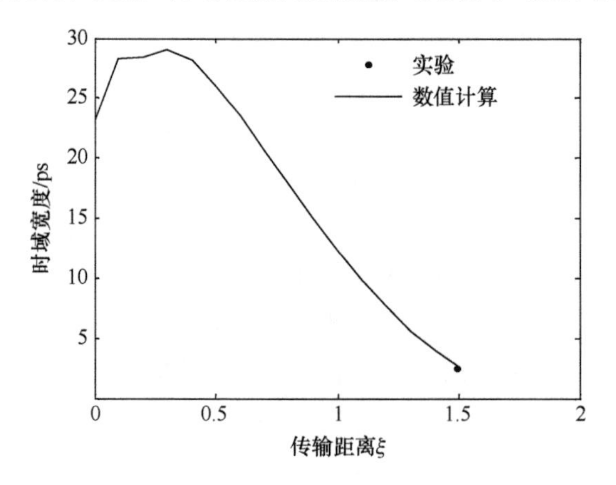

图 5-10　数值计算得到的输入光脉冲时域宽度随归一化传输距离(17.032km)的变化规律

实线所示是数值计算结果，实点所示是 I_p=3A 时输出光脉冲的实验数据

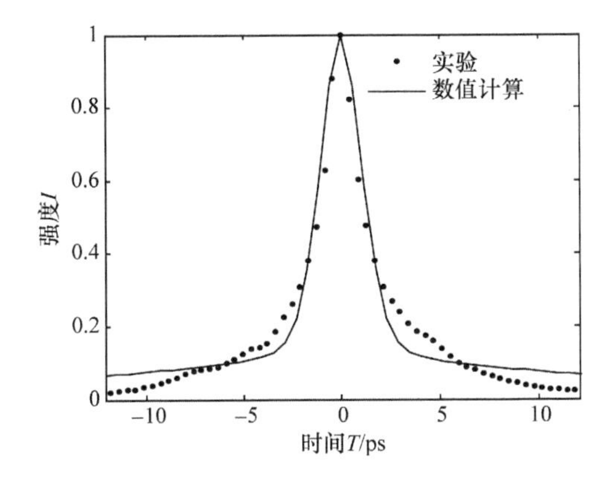

图 5-11　输出脉冲时域宽度 2.42ps 的时域波形

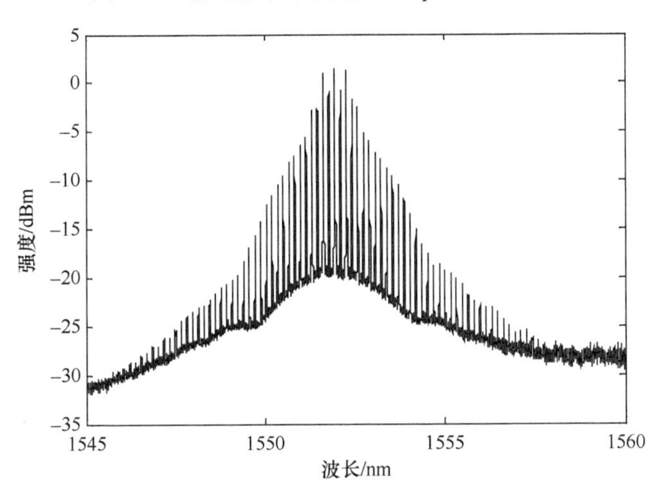

图 5-12　I_p=3A 时输出光脉冲的实验光谱

5.2.3　讨论与总结

文献[25]利用分布拉曼放大在 20km 的色散渐减光纤中将一个 40GHz 拍频信号压缩成 2.2ps 脉冲。Ju Han Lee 等采用高阶孤子压缩效应和非线性开关效应在 8.5km 色散渐减光纤环境中将 12ps 的高斯脉冲压缩成 3ps 的无底座脉冲[26]。文献[27]利用基于 20 km 色散渐减光纤的二极管泵浦拉曼放大器将一个 13ps 的种子脉冲压缩成高质量的 1.3ps 的脉冲。Travers J C 等利用 1.06μm 处孤子压缩效应在色散渐减光子晶体光纤中实现了脉冲压缩比率约为 15 的光脉冲压缩实验[28]。与上述文献[25—28]相比，我们建议的色散渐减光纤中光脉冲的时域压缩比率大约为 10，远大于文献[25]和[26]中的光脉冲压缩比率，与文献[27]的压缩比率近似相等，比文献[28]的光脉冲压缩比率要小。然而，文献[25—28]涉及的色散渐减光纤的制造技术、工艺等非常复杂，色散渐减光纤的制造远比由常规光纤组成的色散渐减光纤的制作困难得多；同时，文献[25—28]在色散渐减光纤中研究光脉冲的压缩采用了不能测量光脉冲时域波形等信息的自相关技术。我们利用二次谐波频率分辨光学门技术能够准确测量光脉冲在色散渐减光纤中的时域波形等参量。采用二次谐波频率分辨光学门技术测量光脉冲在色散渐减光纤中的压缩特性比采用自相关技术测量光脉冲在色散渐减光纤中的压缩特性要好。

综上所述，光脉冲在色散渐减光纤中实现时域压缩，并采用二次谐波频率分辨光学门技术研究压缩特性是一个获得皮秒光脉冲源的好方法。采用二次谐波频率分辨光学门技术实验得到的脉冲压缩数据与数值计算结果吻合。光脉冲在色散渐减光纤中传输时能够有效压缩脉冲时域，光脉冲谱域被展宽。脉冲压缩效应随输入光功率增加而增强的现象与脉冲光谱展宽相关联。

5.3 啁啾光脉冲在变参量系统中传输特性的研究

自从光孤子脉冲被 Hasegawa 和 Tappert[1]理论预测，Mollenauer、Stolen 和 Gordon[35]实验观测到以来，描述光孤子脉冲传输的非线性薛定谔方程由于其科学上的重要性和在等离子物理、凝聚态物理、非线性光学和光通信等领域的重要应用得到世界科技工作者的普遍关注[36-64]。按照此理论，光孤子脉冲可以在无损耗单模光纤中长距离传输，光孤子脉冲在群速色散和克尔效应诱导的自相位调制相互作用的平衡下能保持波形时域不变。著名的非线性薛定谔方程的精确解可以在特定的情况下采用逆散射方法和 Hirota 直接方法求解方程得到[39,40]。在以往的研究中，人们关注较多的是非线性方程相关的特定物理效应，如高阶色散[37, 38, 41]、自陡效应[42-45]、孤子脉冲相互作用[5,46-48]，脉冲内受激拉曼散射[49]和孤子脉冲的稳定传输等[50]；得到了一些非常重要的应用，如光脉冲压缩和孤子脉冲产生[51-58]、全光开关和耦合器[59-61]及光孤子通信[62, 63]等。人们已经认识到光孤子脉冲在非线性科学的基础研究以及下一代高速、大容量、长距离光通信系统中的重要地位。同时，我们注意到上述研究都是基于常量的广义非线性方程或者常量的修改的非线性方程。然而，在实际的光孤子脉冲传输系统中，色散、非线性、增益和损耗通常随系统传输距离的变化而变化。

近年来，光孤子控制或者光孤子管理概念由于其潜在的应用价值得到高度关注和迅速发展。光孤子控制或者光孤子管理系统一般由可变参量非线性方程来描述[14-16, 64-70]。通常情况下，可变参量非线性方程是不可积分的。特定情况下可变参量非线性方程的精确解可以通过构造色散参量 $D(z)$、非线性参量 $R(z)$、衰减与

放大参量 $\Gamma(z)$ 间的特定可积函数关系求解得到[64-70]；然而，上述描述情况难于在实际的系统中实现。本章不去刻意构造上述各个参量间的函数关系和寻求孤子解，而是构建实际可行或接近实际可行的变参量系统，研究常规啁啾光孤子在此可变参量系统中的传输特性，以便于实验实现、应用和服务于社会生活。就我们所知，常规光孤子在实际可行的变参量系统等传输研究中罕有报道，同时考虑到光孤子脉冲通常具有对脉冲传输影响较大的频率啁啾并且频率啁啾可以通过各种方法调节和控制[14-18, 37,38, 58]，本节提出并采用分步傅里叶计算方法研究了啁啾孤子光脉冲在变参量系统等的传输特性。我们主要关注光孤子脉冲的演化和不同情况下的传输稳定性，与文献[64—70]构造非线性方程的可积函数关系并求得方程精确解有着根本不同。

5.3.1　理论模型

当光纤的色散、非线性、增益或损耗沿着光纤长度变化时，如果这种变化对于光纤的模式理论的影响非常小，或者考虑局部模式，使用微扰法，可近似认为光孤子脉冲传输可由变系数的非线性齐次方程描述[64-70]：

$$\mathrm{i}\frac{\partial q}{\partial z}+\frac{1}{2}D(z)\frac{\partial^2 q}{\partial \tau^2}+R(z)|q|^2 q=\mathrm{i}\Gamma(z)q \tag{5-4}$$

方程(5-4)中，$q(z,\tau)$是光孤子脉冲电场的复振幅，z 是归一到色散长度的传输距离，τ是归一化时间，$D(z)$是群速色散系数，$R(z)$是非线性参量，$\Gamma(z)$ 是放大或者损耗参量。按照方程(5-4)，我们可以研究光脉冲在变参量光纤系统中的传输，如光孤子控制或者光孤子管理系统中的光脉冲传输。本节中，我们考虑一个周期分布放大系统[66]，其变群速色散参量为

$$D(z)=\frac{1}{D_0}\exp(\sigma z)R(z) \tag{5-5}$$

变非线性参量为

$$R(z)=R_0+R_1\sin(gz) \tag{5-6}$$

增益或损耗参量为

$$\Gamma(z)=\sigma/2 \tag{5-7}$$

式中，D_0 参量与光脉冲初始峰值功率相关联，R_0、R_1 和 g 为克尔非线性相关参量，σ 是色散渐减或者损耗系统($\sigma<0$)的损耗系数、或者增益放大系统($\sigma>0$)的增益系数。在本节中，令 $D_0=1$，$R_0=0$，$R_1=1$ 和 $g=1$[66]。输入线性啁啾常规光孤子为

$$q(0,\tau)=\operatorname{sech}(\tau)\exp\left(-\frac{\mathrm{i}C\tau^2}{2}\right) \tag{5-8}$$

式中，C 是线性啁啾参量。为了减弱计算窗口的边缘效应，我们设置了较大的时域计算窗口 τ =(–320，320)和取样数目 4096。

5.3.2 数值研究

5.3.2.1 光孤子脉冲时域宽度随传输距离的变化

图 5-13 所示是光孤子脉冲时域半峰全宽随归一化传输距离的变化规律。图 5-13(a)中实线、虚线、点划线和点线所示分别对应无啁啾孤子脉冲 σ =－0.05，－0.025，0.025 和 0.05 的情况，图 5-13(b)中曲线 1～5 所示是 σ =－0.05 时分别对应啁啾孤子脉冲 C=－0.2，－0.1，0，0.1 和 0.2 的情况；图 5-13(c)中曲线 1～5 所示是 σ = 0 时分别对应啁啾孤子脉冲 C=－0.2，－0.1，0，0.1 和 0.2 的情况，图 5-13(d)中曲线 1～5 所示是 σ =0.05 时分别对应啁啾孤子脉冲 C=–0.2，–0.1，0，0.1 和 0.2 的情况。由图 5-13(a)可以得到，无啁啾光孤子脉冲时域半峰全宽随传输距离增加产生轻微振荡，振荡幅度随增益或损耗参量 σ 的增加逐渐增大，当 σ<0 时，光孤子脉冲时域逐渐展宽。由图 5-13(b)～(d)可以得到，啁啾光孤子脉冲在损耗系统(b)、无损耗系统(c)和增益放大系统(d)中随传输距离的增加作不同程度的周期性压缩和展宽。系统参量|C|和 σ 越大，光脉冲时域宽度随传输距离的变化越大。负啁啾使非线性减弱，使光脉冲时域展宽；正啁啾使非线性加强，使光脉冲时域压缩。啁啾对周期放大系统中的光脉冲时域宽度的影响比对损耗系统中光脉冲宽度的影响要大。

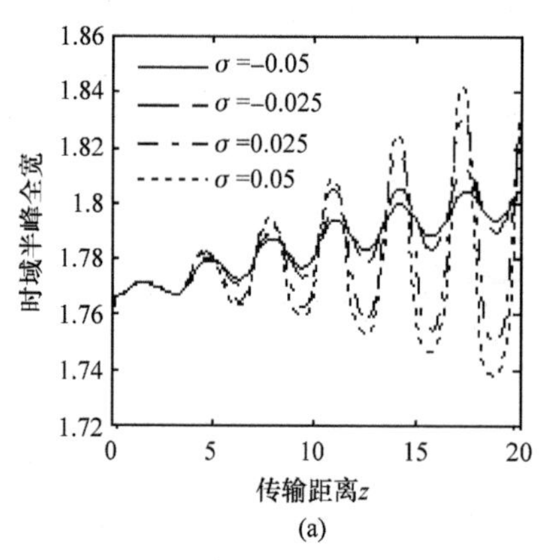

(a)

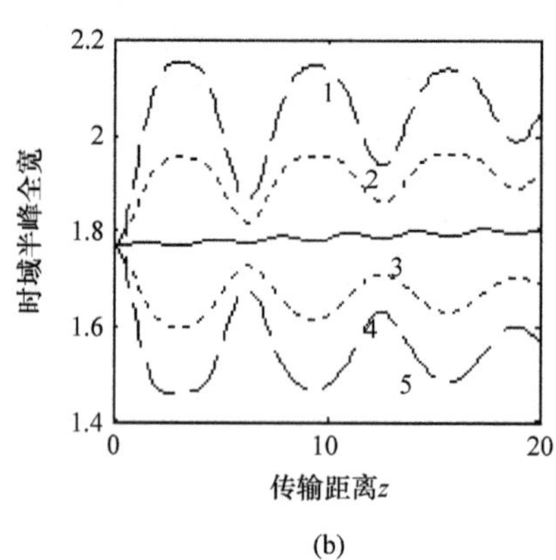

(b)

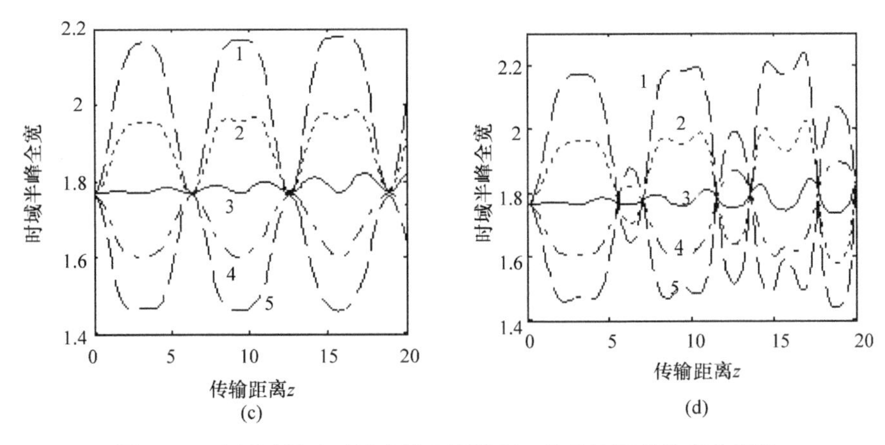

图 5-13　光孤子脉冲时域半峰全宽随归一化传输距离的变化规律

(b)曲线 1～5 所示是σ= –0.05 时分别对应啁啾孤子脉冲 C= –0.2，–0.1，0，0.1 和 0.2 的情况；(c)曲线 1～5 所示是σ=0 时分别对应啁啾孤子脉冲 C= –0.2，–0.1，0，0.1 和 0.2 的情况；(d)曲线 1～5 所示是σ=0.05 时分别对应啁啾孤子脉冲 C= –0.2，–0.1，0，0.1 和 0.2 的情况

5.3.2.2　时域波形的演化

图 5-14 所示是当σ =0.05 时，C=0(a)和– 0.2(b)对应的时域波形随传输距离的变化规律，以及 z =20 处 C=0(c)和–0.2(d)的归一化时域波形。图 5-14(c)和(d)中实线、虚线、点线和点划线分别是数值计算结果、指数曲线、高斯脉冲曲线和双曲正割脉冲曲线。由图 5-14(a)可以得到，无啁啾光孤子脉冲瞬变电场$|q(z, \tau)|$在放大系统(σ = 0.05)中能够逐渐得到增益，实现放大。时域波形的峰值和相应的时域宽度随传输距离稍有变化。由图 5-14(b)可以得到，啁啾光孤子脉冲峰值在放大系统中产生了明显的振荡，啁啾参量$|C|$越大，脉冲峰值变化越明显。由图 5-14(c)和(d)可以得到，光孤子脉冲传输 20 个色散长度(约为 12.74 个孤子周期)后的时域波形仍与双曲正割曲线吻合，表明传输后光脉冲仍然维持孤子特性。

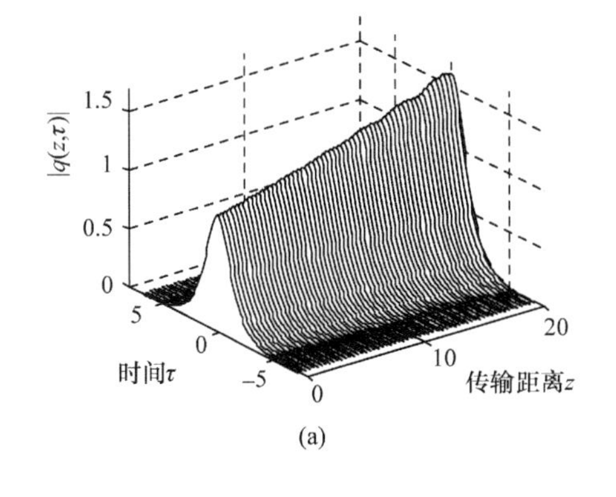

(a)

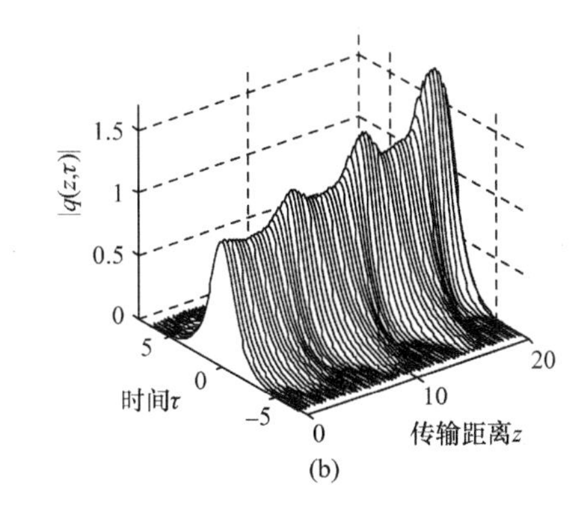

(b)

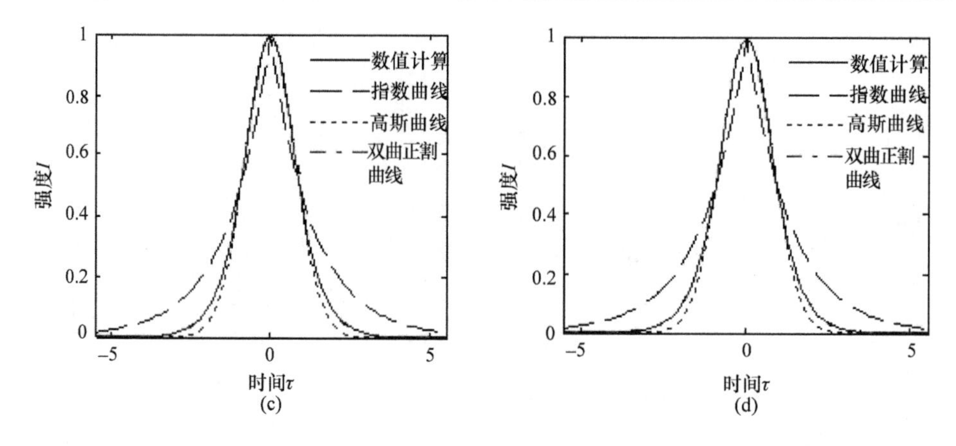

图 5-14　当σ=0.05 时 C=0 (a)和−0.2(b)对应的时域波形随传输距离的变化规律，以及 z=20 处 C=0 (c)和−0.2(d)的归一化时域波形

5.3.2.3　光脉冲光谱的演化

图 5-15 所示是当$\sigma = 0.05$ 时 C=0(a)和− 0.2(b)对应的光孤子脉冲光谱随传输距离的变化规律。

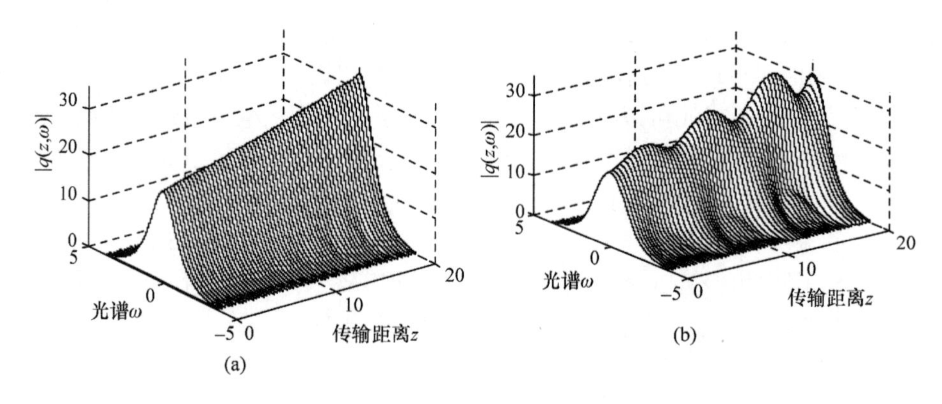

图 5-15　当σ=0.05 时 C=0(a)和−0.2(b)对应的光孤子脉冲光谱随传输距离的变化规律

可以得到，无啁啾光孤子脉冲光谱$|q(z, \omega)|$在放大系统($\sigma = 0.05$)中随传输距离增加能够线性放大，在损耗系统($\sigma = -0.05$)中随传输距离增加能够线性变小。在啁啾情况下，光孤子脉冲光谱在放大系统中产生了明显的振荡，啁啾参量$|C|$越大，脉冲光谱的振荡变化越明显，光脉冲光谱仍然维持单峰结构。

5.3.2.4　光孤子脉冲传输的稳定性

下面研究光孤子脉冲在小增益或者低损耗系统($\sigma = 0.01$)、考虑实际拉曼放大和光纤衰减的微扰系统中的传输稳定性。为了展示光孤子脉冲的传输稳定性，我们给出了三种类型的数值实验，如图 5-16 所示。图 5-16(a)所示是光孤子脉冲时域波形在最大幅值为 0.1 的随机微扰情况下的演化规律，(b)所示是光孤子脉冲时域波形在初始幅值 10%微扰情况下的演化规律，(c)所示是光孤子脉冲时域波形在 0.01sin$z$ 增益或损耗微扰情况下的演化规律，(d)所示是光孤子脉冲在 0.01sin$z$ 增益或损耗微扰情况下传输 100 个色散长度处时域波形。由图 5-16 可以得到，光孤子脉冲在不同微扰光纤系统中能够稳定传输 100 个色散长度，表明三种微扰均不能影响光孤子脉冲传输的稳定性，除了使光脉冲产生轻微的扰动外。

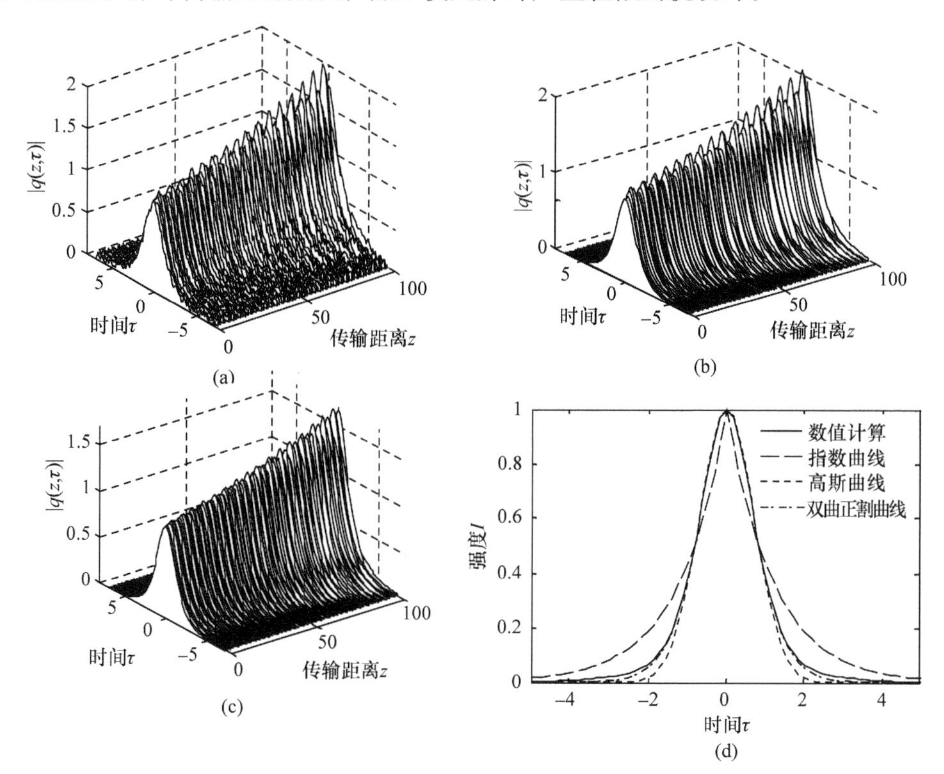

图 5-16　(a)光孤子脉冲时域波形在最大幅值为 0.1 的随机微扰情况下的演化规律，(b)光孤子脉冲时域波形在初始幅值 10%微扰情况下的演化规律，(c)光孤子脉冲时域波形在 0.01sin$z$ 增益或损耗微扰情况下的演化规律，(d)光孤子脉冲在 0.01sin$z$ 增益或损耗微扰情况下传输 100 个色散长度处的时域波形

5.3.3　讨论与小结

光孤子脉冲在孤子控制系统中传输仍能保持孤子特性。输出光孤子脉冲光谱仍呈单峰结构。无啁啾光孤子脉冲的时域宽度随传输距离的增加产生轻微振荡，

振荡幅度随增益或损耗参量σ的增加而增加。啁啾孤子脉冲随传输距离的增加产生明显地周期性压缩和展宽，线性啁啾参量$|C|$和损耗参量σ越大，脉冲时域宽度随距离的变化就越大。啁啾对周期放大系统中的光脉冲时域宽度的影响比对损耗系统中光脉冲宽度的影响要大。三种数值实验表明，在不同微扰下，孤子脉冲能够在周期放大系统中稳定传输。实际应用中，我们通过控制和调节变参量周期分布放大系统中的参量可以获得合适的孤子光脉冲。

5.4　啁啾脉冲的碰撞特性

双折射光纤中的孤子碰撞具有科学上的重要性，在光逻辑器件和偏振复用系统中具有重要的实用价值，从而引起了科技工作者的广泛兴趣[19, 71-82]。两个偏振孤子的初始间隔可以根据实际应用灵活设置。在光逻辑器件中，一个偏振态孤子作为数据信号，另一个偏振态孤子作为控制信号，其初始时间间隔可以不同；在输入端，通过调节控制信号孤子和数据信号孤子的时间间隔、啁啾等来控制两孤子是否处于束缚态，使其产生不同的时间延迟；在输出端可以通过设置相应的延迟来接收或不接收数据信号。在偏振复用孤子通信中两个偏振态的比特流是各自独立的，在时域上相互交错，相对初始时间间隔可长可短。孤子束缚态可导致系统更高的通信比特率。Menyuk C R 推导了无啁啾孤子脉冲在双折射光纤中的非线性耦合方程，并研究了无啁啾孤子脉冲的传输稳定性[19, 76-78]。随后，无啁啾正交偏振孤子碰撞特性得到了普遍关注[79-82]。考虑到脉冲通常具有较大的频率啁啾且可以通过传输长度、采用啁啾光栅技术或预啁啾技术等进行调节[14-16, 19]，本节数值研究了在双折射光纤中啁啾正交偏振孤子间的碰撞特性，对光逻辑器件和偏振复用进一步研究有重要的指导作用。

5.4.1　双折射光纤中孤子碰撞的理论模型

双折射光纤中的孤子传输可以用耦合方程(1-22)来描述，考虑到双折射光纤中的孤子碰撞特性研究主要应用于光逻辑器件中，传输距离短，故忽略高阶色散和光纤损耗，方程(1-22)修正为

$$\mathrm{i}\left(\frac{\partial u}{\partial \xi}+\delta\frac{\partial u}{\partial \tau}\right)+\frac{1}{2}\frac{\partial^2 u}{\partial \tau^2}+(|u|^2+B|v|^2)u=0 \tag{5-9a}$$

$$\mathrm{i}\left(\frac{\partial v}{\partial \xi}-\delta\frac{\partial v}{\partial \tau}\right)+\frac{1}{2}\frac{\partial^2 v}{\partial \tau^2}+(|v|^2+B|u|^2)v=0 \tag{5-9b}$$

式中，u 和 v 分别是正交偏振孤子脉冲的归一化慢变包络电场；归一化时间

$\tau=(t-\overline{\beta}_{1z})/T_0$，归一化到脉冲的半宽度 T_0；$\overline{\beta}_1=0.5(\beta_{1x}+\beta_{1y})$ 与脉冲的平均群速度成反比；$\delta=(\beta_{1x}-\beta_{1y})T_0/2|\beta_2|$，反映了两个脉冲的群速度失配；$\delta$>0 时，$u$ 和 v 分别是慢孤子和快孤子；$0\leqslant B\leqslant 1$ 是交叉相位调制系数；B=0 时，方程(5-9)简化为非耦合的非线性薛定谔方程[19]；B=1 时，方程(5-9)简化为具有解析解的 Manakov 方程[19, 83-87]；B=2/3 时对应线性双折射情况，方程(5-9)不能解析求解[19]。作者采用分步傅里叶方法，根据方程(5-9)数值研究了啁啾孤子在线性双折射光纤中的碰撞特性，两个正交偏振的输入孤子脉冲分别是

$$u(\xi=0,\tau)=\operatorname{sech}(\tau+\tau_0)\exp\left[-\frac{\mathrm{i}C(\tau+\tau_0)^2}{2}\right] \tag{5-10a}$$

和

$$v(\xi=0,\tau)=\operatorname{sech}(\tau-\tau_0)\exp\left[-\frac{\mathrm{i}C(\tau-\tau_0)^2}{2}\right] \tag{5-10b}$$

式中，C 为线性啁啾参量，$2\tau_0$ 为两孤子时间间隔。根据分步傅里叶算法，当脉冲从 ξ 传输到 $\xi+h$(微小距离 h)时，色散效应和非线性效应可以分别单独计算，使脉冲在光纤中的传输问题变得简单。为了减小边缘效应，作者设置较大的时域计算窗口 $\tau=(-80,80)$ 和较大的采样点数 512。

5.4.2　初始线性啁啾对孤子碰撞特性的影响

5.4.2.1　啁啾对孤子时间间隔变化的影响

两孤子时间间隔定义为

$$\Delta\tau(\xi)=\tau_{\max}^{u}(\xi)-\tau_{\max}^{v}(\xi) \tag{5-11}$$

式中，$\tau_{\max}^{u}(\xi)$ 和 $\tau_{\max}^{v}(\xi)$ 分别是 $|u(\xi,\tau)|$ 和 $|v(\xi,\tau)|$ 最大值对应的时间位置。由方程(5-9)和(5-10)可以得到 $\tau_{\max}^{u}(\xi)=-\tau_{\max}^{v}(\xi)$。当 $|\Delta\tau(\xi)|$<0 时，慢孤子在快孤子之前；当 $|\Delta\tau(\xi)|$ 随传输距离增加而不断增大时，两孤子将逐渐分离；若 $\Delta\tau(\xi)$ 随传输距离的变化越来越小，最终随传输距离的增加不变，此时的 δ 称为孤子束缚态的阈值 δ_{th}。当 $\delta\leqslant\delta_{\mathrm{th}}$ 时，两正交偏振孤子在传输过程中相互束缚在一起；当 $\delta>\delta_{\mathrm{th}}$ 时，两孤子相互分离。双折射光纤中孤子束缚态的阈值 δ_{th} 由孤子时间间隔随传输距离的变化决定。初始入射时，设置两孤子脉冲部分分离，即 $2\tau_0=1.25$。图 5-17 所示是 $2\tau_0=1.25$ 时孤子时间间隔随传输距离的变化。横坐标是传输距离，归一化到孤子周期；纵坐标是孤子时间间隔，归一化到孤子半宽度 T_0。虚线所示对应 $\delta=0.832$ 和 $C=0.5$ 的情况，实线对应 $\delta=0.615$ 和 $C=0$ 的情况，点线对应 $\delta=0.46$ 和 $C=-0.5$ 的情况。由图 5-17 可以得到，$C=0$ 时

$\delta_{\text{th}}=0.615$, $C=0.5$ 时 $\delta_{\text{th}}=0.832$, $C=-0.5$ 时 $\delta_{\text{th}}=0.46$。这表明初始啁啾改变了双折射光纤中的孤子束缚态阈值 δ_{th}，正啁啾使阈值 δ_{th} 增大，负啁啾使阈值 δ_{th} 减小。这对光逻辑器件的应用非常重要。

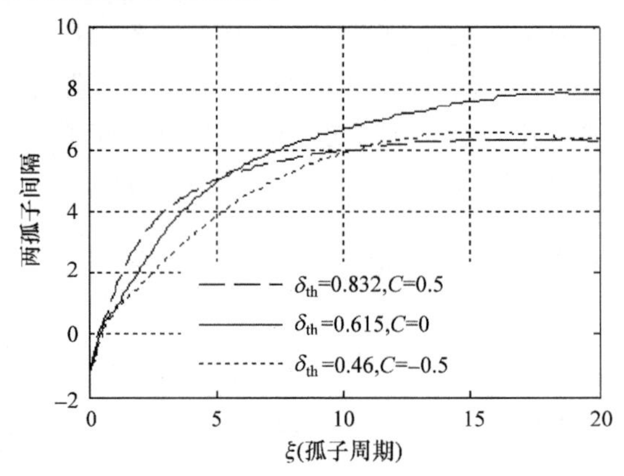

图 5-17　$2\tau_0=1.25$ 时啁啾孤子时间间隔随传输距离的变化

5.4.2.2　啁啾对孤子混合演化的影响

描述孤子碰撞的孤子混合参量 $m(\xi)$ 定义为

$$m(\xi)=\frac{\int_{-\infty}^{+\infty}|u(\xi,\tau)||v(\xi,\tau)|\,\mathrm{d}\tau}{\int_{-\infty}^{+\infty}|u(\xi=0,\tau)|^2\,\mathrm{d}\tau} \tag{5-12}$$

孤子混合参量 $m(\xi)$ 能够表示两孤子的混合程度，其数值在 0～1 变化。当 $m(\xi)=1$ 时，两相同孤子完全重叠；当 $m(\xi)=0$ 时，两孤子完全分离；当 $m(\xi)$ 接近 1 时，两孤子相互捕获形成孤子束缚态。图 5-18 所示是 $\delta=0.7$ 和 $2\tau_0=1.25$ 时线性啁啾孤子的混合参量 $m(\xi)$ 随归一化传输距离的变化，曲线 1～5 分别对应 $C=-0.5$、-0.3、0、0.3 和 0.5 的情况。横坐标是传输距离，归一化到孤子周期，纵坐标是孤子混合参量 $m(\xi)$。由曲线 4 可得，当 $C=0.3$ 时，两初始输入孤子在 $\xi=0$ 处部分分离且 $m=0.78$；在 $\xi\approx0.5$ 处快孤子赶上慢孤子且 $m\approx1$；接着，两孤子在 $\xi\approx1$ 处再次部分分离且 $m\approx0.6$；后随初始距离的增加 m 逐渐增大，两啁啾孤子相互捕获对方的小部分能量；最终，两孤子在 $\xi>4$ 处形成孤子束缚态且 $m\approx0.9$。

在 $0.2\leqslant C\leqslant1$ 时，两线性啁啾孤子随初始距离的增加(约 $\xi>4$ 处)最终形成孤子束缚态；啁啾 C 值越大，形成束缚态的距离越短。在 $-1\leqslant C<0.2$ 时，两孤子随初始距离的增加不能形成孤子束缚态。

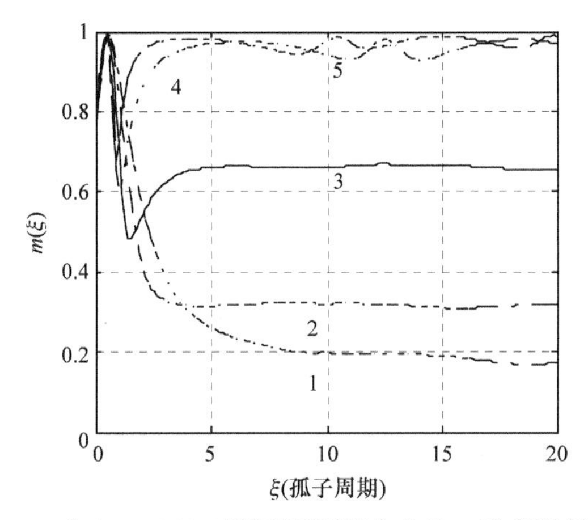

图 5-18　$\delta=0.7$ 和 $2\tau_0=1.25$ 时啁啾孤子混合参量 $m(\xi)$ 随传输距离的变化

曲线 1～5 分别对应 $C=-0.5$、-0.3、0、0.3 和 0.5

5.4.2.3　啁啾对频率质心变化的影响

两孤子的频率质心定义为

$$\omega_{\text{cent}}^{u}(\xi)=\frac{\int_{-\infty}^{+\infty}\omega\left|u(\xi,\omega)\right|^{2}\mathrm{d}\omega}{\int_{-\infty}^{+\infty}\left|u(\xi,\omega)\right|^{2}\mathrm{d}\omega} \tag{5-13a}$$

$$\omega_{\text{cent}}^{v}(\xi)=\frac{\int_{-\infty}^{+\infty}\omega\left|v(\xi,\omega)\right|^{2}\mathrm{d}\omega}{\int_{-\infty}^{+\infty}\left|v(\xi,\omega)\right|^{2}\mathrm{d}\omega} \tag{5-13b}$$

由方程(5-9)、(5-10)和(5-13)可以得到 $\omega_{\text{cent}}^{u}(\xi)=-\,\omega_{\text{cent}}^{v}(\xi)$。图 5-19 所示是 $\delta=0.7$ 和 $2\tau_0=1.25$ 时线性啁啾孤子频率质心 $\omega_{\text{cent}}^{u}(\xi)$ 和 $\omega_{\text{cent}}^{v}(\xi)$ 随归一化距离的变化。实线 1～5 所示分别是对应 $C=-0.5$、-0.3、0、0.3 和 0.5 时的 $\omega_{\text{cent}}^{u}(\xi)$，虚线 6～10 所示分别是对应 $C=-0.5$、-0.3、0、0.3 和 0.5 时的 $\omega_{\text{cent}}^{v}(\xi)$，横坐标是归一化传输距离，纵坐标是孤子的频率质心。由图 5-19 可得，在双折射效应和非线性效应共同作用下，慢孤子的频率质心在碰撞前红移，在碰撞后蓝移。快孤子的频率质心 $\omega_{\text{cent}}^{v}(\xi)$ 随传输距离的变化与慢孤子的变化相反。这改变了孤子群速度，从而使得慢孤子变快和快孤子变慢，最终导致 $0.2\leqslant C\leqslant 1$ 时两啁啾孤子以相同的速度传输。当 $-1\leqslant C<0.2$ 时，频移量随啁啾的增加而增大；当 $0.2\leqslant C\leqslant 1$ 时，频移量随啁啾和距离的增加出现了明显的振荡结构。

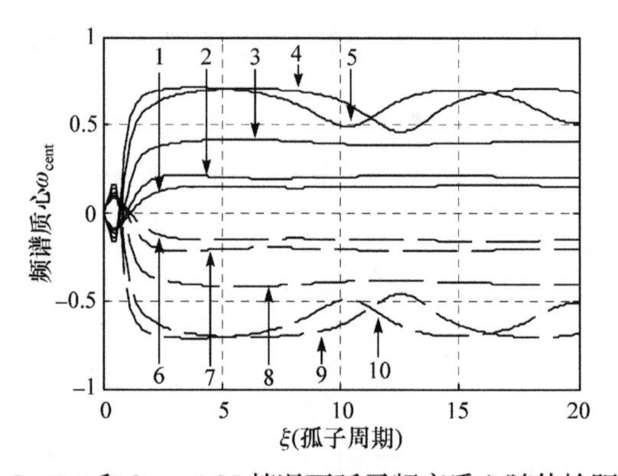

图 5-19　$\delta=0.7$ 和 $2\tau_0=1.25$ 情况下孤子频率质心随传输距离的变化

实线 1～5 分别对应 C= −0.5、−0.3、0、0.3 和 0.5 时的 $\omega_{\text{cent}}^{u}(\xi)$，虚线 6～10 分别对应 C= −0.5、−0.3、0、0.3 和 0.5 时的 $\omega_{\text{cent}}^{v}(\xi)$

5.4.2.4　啁啾对时域半峰全宽变化的影响

T_{FWHM}^{u} 和 T_{FWHM}^{v} 分别是慢、快孤子的时域半峰全宽。由方程(5-9)和方程(5-10)可得 $T_{\text{FWHM}}^{u}=T_{\text{FWHM}}^{v}$。图 5-20 所示是 $\delta=0.7$ 和 $2\tau_0=1.25$ 时线性啁啾慢孤子半峰全宽 T_{FWHM}^{u} 随归一化距离的变化，横坐标是归一化传输距离，纵坐标是孤子时域半峰全宽，归一化到 T_0，实线 1～5 所示分别对应 $C=-1$、−0.5、0、0.5、1的情况。数值结果发现，啁啾孤子时域半峰全宽随传输距离的增加发生衰减振荡。

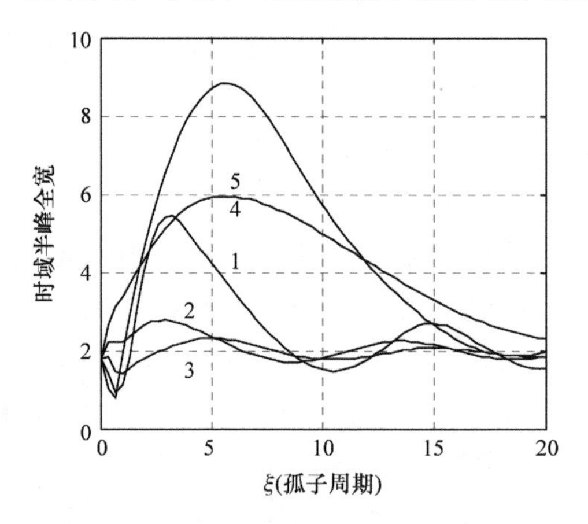

图 5-20　$\delta=0.7$ 和 $2\tau_0=1.25$ 时啁啾孤子时域半峰全宽随传输距离的变化

实线 1～5 分别对应 C= −1、−0.5、0、0.5 和 1

在 $-1\leqslant C\leqslant 0.2$ 和 $0.6\leqslant C\leqslant 1$ 情况下，振荡周期和振幅随啁啾的增加而增加；在$-0.2<C<0.6$ 情况下，振荡周期和振幅随啁啾的增加而减小。给定相同的

啁啾|C|值，正啁啾对应的振荡幅度大于负啁啾的情况，这表明正啁啾对啁啾孤子时域半峰全宽的影响比负啁啾的要大。在非线性效应和啁啾共同作用下，孤子在正啁啾时于 $\xi=1$ 附近出现脉冲压缩。

5.4.2.5　啁啾对时域波形演化的影响

由方程(5-9)和(5-10)可得 $u(\xi,\tau)=v(\xi,-\tau)$。图 5-21 所示是 $C=0.5$、$\delta=0.7$ 和 $2\tau_0=1.25$ 时慢孤子 $u(\xi,\tau)$ 时域波形随传输距离的变化，x 坐标是归一化时间，y 坐标是归一化传输距离，z 坐标是孤子归一化振幅。快孤子 $v(\xi,\tau)$ 时域波形随传输距离的变化是慢孤子情况关于 $\tau=0$ 的镜像。数值结果发现，$C=0.5$ 时，两啁啾孤子初始输入时部分分离，在传输过程中孤子脉冲首先进行初始压缩，碰撞后逐渐形成孤子束缚态。在非线性效应和初始啁啾作用下，脉冲边缘产生明显的色散波。时域波形峰值随初始距离的增加而产生振荡，啁啾情况下的振荡周期和振幅比无啁啾时的大，并且随啁啾|C|的增加而变化。这样，输出脉冲时域波形的峰值可以通过调节初始啁啾大小来控制，这对光逻辑器件的应用具有一定的指导意义。由方程(5-9)的不可积性可推知色散波辐射的必然存在，啁啾|C|越大，啁啾孤子脉冲边缘的色散波越强。

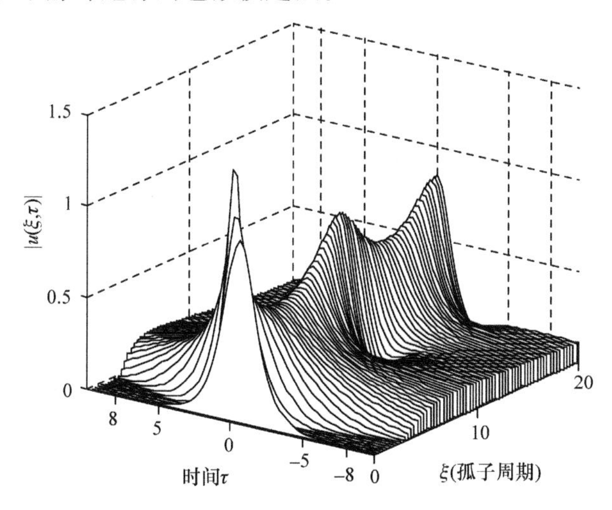

图 5-21　$\delta=0.7$、$2\tau_0=1.25$ 和 C=0.5 时孤子 $|u(\xi,\tau)|$ 时域波形随传输距离的演化

5.4.2.6　啁啾对 δ_{th} 和 $2\tau_0$ 之间关系的影响

研究过程中考虑了 $|2\tau_0|\geqslant 5$ 时两孤子完全分离的情况；$2\tau_0<-5$ 时快孤子在前，慢孤子在后，两孤子相互作用可以忽略；$2\tau_0>5$ 时慢孤子在前，快孤子在后，其相互作用与 $2\tau_0=5$ 时类似。图 5-22 所示是啁啾对 δ_{th} 和 $2\tau_0$ 之间关系的影

响。实线所示对应 $C=0$ 时的关系曲线，虚线所示对应 $C=0.5$ 的情况，点线所示对应 $C=-0.5$ 的情况。横坐标是孤子时间间隔，归一化为 $2\tau_0$；纵坐标是群速失配系数 δ。由图 5-22 可得，孤子束缚态阈值 δ_{th} 取决于孤子的初始啁啾和时间间隔。在 $C=0$ 情况下，$|2\tau_0|\leqslant 3$ 时，阈值 δ_{th} 随初始时间间隔的增加而明显减小；最大值对应 $\tau_0=0$；$\tau_0>3$ 时，几乎保持不变。这种变化与文献[80—82]的结果一致。在 $C=0.5$ 情况下，$-1<2\tau_0\leqslant 2.1$ 时，阈值 δ_{th} 明显比无啁啾情况时大；$2.1<2\tau_0\leqslant 5$ 时，阈值 δ_{th} 明显比无啁啾情况时小；$-4<2\tau_0\leqslant -1$ 时，阈值 δ_{th} 比无啁啾情况时稍小。在 $C=-0.5$ 情况下，$-5\leqslant 2\tau_0<0$ 时，阈值 δ_{th} 比无啁啾情况时稍大；$0\leqslant 2\tau_0\leqslant 5$ 时，阈值 δ_{th} 明显比无啁啾情况时小。阈值 δ_{th} 的最大值在 $C>0$ 时移向 $\tau_0>0$ 区域，在 $C<0$ 时移向 $\tau_0<0$ 区域。

研究表明初始啁啾改变了孤子束缚态的阈值 δ_{th}，初始正啁啾对阈值变化的影响比负啁啾的大。因此，光逻辑器件的设计必须考虑脉冲初始啁啾的影响。

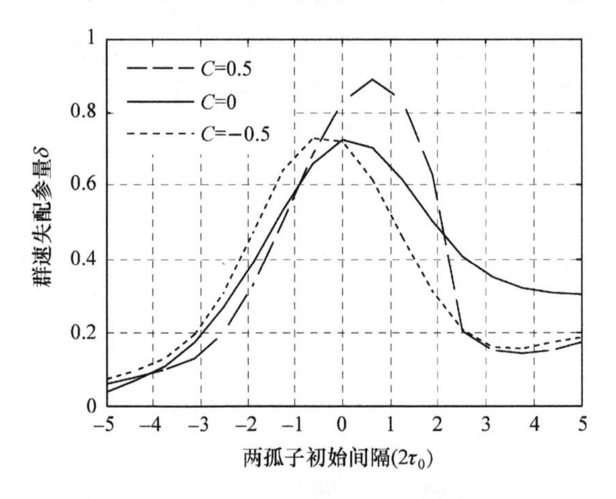

图 5-22　δ_{th} 和 $2\tau_0$ 的关系曲线

5.4.3 讨论与小结

本节采用分步傅里叶方法数值研究了具有初始线性啁啾的正交极化孤子脉冲在线性双折射光纤中的碰撞特性。研究表明，初始啁啾改变了光孤子在双折射光纤中形成束缚态的阈值 δ_{th}，正啁啾对光孤子束缚态阈值 δ_{th} 的影响比负啁啾明显。因此，可以改变脉冲的初始啁啾来控制输出脉冲幅度。

5.5 本章小结

本章利用二次谐波频率分辨光学门(SHG-FROG)脉冲分析仪实验研究了具有较大啁啾光脉冲在色散平坦光纤正常色散区的传输规律和特点[88]，实验结果表明：①输出脉冲的时域宽度随着输入光脉冲平均功率的增加而减小；②光纤色散越小，输出脉冲宽度越窄；③负啁啾光脉冲传输后演化成近高斯脉冲；④输出谱宽随输入光功率增加而增加。

本章提出了光脉冲在色散渐减光纤中实现时域压缩以获得 ps 量级光脉冲源的好方法；采用 SHG-FROG 技术实验研究得到的脉冲压缩数据与数值计算结果吻合[89]。光脉冲在色散渐减光纤中能够有效压缩，光脉冲谱域被展宽。

本章采用分步傅里叶方法数值研究了变参量周期分布放大系统中的脉冲传输；研究表明，光孤子脉冲在孤子控制系统中传输仍能保持孤子特性；啁啾脉冲传输中产生明显地周期性压缩和展宽，线性啁啾参量$|C|$和损耗参量σ越大，脉冲时域宽度越大。三种微扰数值实验表明，在周期放大系统中，孤子脉冲在不同微扰下能够稳定传输；实际应用中，我们通过控制和调节系统参量可以获得合适的孤子光脉冲[90]。

此外，采用分步傅里叶方法数值研究了啁啾孤子脉冲在线性双折射光纤中的碰撞特性；研究表明，初始啁啾改变了孤子在双折射光纤中形成束缚态的阈值，正啁啾对光孤子束缚态阈值的影响比负啁啾明显。因此可以改变脉冲的初始啁啾来控制输出脉冲幅度[91,92]。

本章研究结果为我们利用啁啾脉冲实现短脉冲源及其传输、啁啾脉冲变参量传输、光逻辑器件等的优化设计提供了重要依据。

参考文献

[1] Hasegawa A, Tappert F. Transmission of stationary nonlinear optical pulses in dispersive dielectric fibers. Ⅰ. Anomalous dispersion. Appl. Phys. Lett., 1973, 23(3): 142～144.

[2] Desaix M, Helczynski L, Anderson D, et al. Propagation properties of chirped soliton pulses in optical nonlinear Kerr media. Phys. Rev. E, 2002, 65: 056602-1.

[3] Li Z H, Li L, Tian H P, et al. Chirped femtosecond soliton like laser pulse form with self-frequency shift. Phys. Rev. Lett., 2002, 89: 263901-1.

[4] Zheng H J, Liu S L. Effects of initial frequency chirp on the linear propagation characteristics of the exponential optical pulse. Chinese Physics, 2006, 15: 1831～1837.

[5] Zheng H J, Liu S L, Xu J P. Effect of initial linear chirp on collision characteristics of two solitons in the birefringent fiber. Chinese Physics, 2007, 16(7)：2023～2027.

[6] Agrawal G P. Nonlinear Fiber Optics. 4th edition. Singapore: Elsevier Pte Ltd., 2009.

[7] Yu J J, Zhang X G, Yang Q M, et al. Normal dispersion fiber elimination chirp to transmit soliton. Study on Optical Communicatons , 1996, 80: 6～9 (in Chinese).

[8] Yu J J, Yang B J, Guan K J. Chirp pulse propagation in normal dispersion fiber. Journal of Beijing University of Posts and Telecommunications, 1997, 20: 59～63 (in Chinese).

[9] Wu J, Lou C Y, Yao M Y, et al. Compression of optical pulses from gain-switched DFB laser diode through normal dispersion fiber. Journal of Tsinghua University(Science and Technology), 1997, 37: 76～79 (in Chinese).

[10] Xia G, Huang D X, Yuan X H. Investigation of supercontinuum generation in normal dispersion-flattened fiber by picosecond seed pulses. Acta Physica Sinica, 2007, 56: 2212～2217 (in Chinese).

[11] Xu Y Z, Ren X M, Wang Z N, et al. Flat supercontinuum generation in a microstructure fiber with normal dispersion. Journal of Optoelectronics · Laser. 2007, 18: 889～892 (in Chinese).

[12] Kieu K, Wise F W. All-fiber normal-dispersion femtosecond laser. Opt. Express, 2008, 16: 11453～11458.

[13] Barry L P, Delburgo S, Thomsen B C, et al. Optimization of optical data transmitters for 40-Gb/s lightwave systems using frequency resolved optical gating. Photon. Tech. Lett., 2002, 14: 971～974.

[14] Zheng H J, Liu S L, Li X, et al. Temporal characteristics of an optical soliton with distributed Raman amplification. J. Appl. Phys, 2007, 102(10): 103106-1～4.

[15] Liu S L, Zheng H J. Measurement of nonlinear coefficient of optical fiber based on small chirped soliton transmission. Chinese Optics Letters, 2008, 6: 533～535.

[16] Zheng H J, Liu S L, Tian Z, et al. Effects of Raman amplification on propagation characteristics of the soliton. Chinese Journal of Lasers, 2008, 35(6): 861～866(in Chinese).

[17] Zheng H J, Liu S L, Li X, et al. Pulse compression in dispersion-decreasing-like fibers. Optics & Laser Technology, 2011, 43(7): 1321～1324.

[18] Barry L P, Dudley J M, Bollond P G, et al. Complete characterisation of pulse propagation in optical fibres using frequency-resolved optical gating. Electronics Letters, 1996, 32 (25): 2339～2340.

[19] Agrawal G P. Nonlinear Fiber Optics. 5th edition. Singapore: Elsevier Pte Ltd., 2012.

[20] Canova F, Uteza O, Chambaret J P, et al. High-efficiency, broad band, high-damage threshold high-index gratings for femtosecond pulse compression. Optics Express, 2007, 15(23): 15324～15334.

[21] Cotel1 A, Castaing M, Pichon P, et al. Phased-array grating compression for high-energy chirped pulse amplification lasers. Opt. Express, 2007, 15(5): 2742～2752.

[22] Mollenauer L F, Stolen R H, Gordon J P, et al. Extreme picosecond pulse narrowing by means of soliton effect in single-mode optical fibers. Opt. Lett., 1983, 8 (5): 289～291.

[23] Murphy T E. 10-GHz 1.3-ps pulse generation using chirped soliton compression in a raman gain medium. IEEE Photonics Technology Letters, 2002, 14(10): 1424～1426.

[24] Iwatsuki K, Suzuki K I, Nishi S. Adiabatic soliton compression of gain-switched DFB-LD Pulse by distributed fiber raman amplification. IEEE Photonics Technology Letters, 1991, 3(12):

1074～1076.

[25] Lee J H, Han Y G, Lee S B. 40 GHz adiabatic compression of a modulator based dual frequency beat signal using Raman amplification in dispersion decreasing fiber. Opt. Express, 2004, 12(10): 2187～2192.

[26] Lee J H, Kogure T, Richardson D J. Wavelength tunable 10-GHz 3-ps pulse source using a dispersion decreasing fiber-based nonlinear optical loop mirror. IEEE Journal of Selected Topics in Quantum Electronics, 2004, 10(1): 181～185.

[27] Kogure T, Lee J H, Richardson D J. Wavelength and duration-tunable 10-GHz 1.3-ps pulse source using dispersion decreasing fiber-based distributed raman amplification. IEEE Photonics Technology Letters, 2004, 16(4): 1167～1169.

[28] Travers J C, Stone J M, Rulkov A B, et al. Optical pulse compression in dispersion decreasing photonic crystal fiber. Opt. Express, 2007, 15(20): 13203～13211.

[29] Druon F, Georges P. Pulse-compression down to 20 fs using a photonic crystal fiber seeded by a diode-pumped Yb: SYS laser at 1070 nm. Opt Express, 2004, 12(15): 3383～3396.

[30] Ouzounov D, Hensley C, Gaeta A, et al. Soliton pulse compression in photonic band-gap fibers. Opt Express, 2005, 13(16): 6153～6159.

[31] Han M, Lou C Y, Wu Y, et al. Generation of pedestal-free 10GHz pulses from a comblike dispersion profiled fiber compressor and its application in supercontinuum generation. Chinese Physics Letters, 2000, 17(11): 806～808.

[32] Inoue T, Tobioka H, Igarashi K, et al. Optical pulse compression based on stationary rescaled pulse propagation in a comblike profiled fiber. Journal of Lightwave Technology, 2006, 24 (7): 2510～2522.

[33] Chernikov S V, Taylor J R, Kashyap R. Experimental demonstration of step-like dispersion profiling in optical fibre for soliton pulse generation and compression. Electronics Letters, 1994, 30(5): 433～435.

[34] Barry L P, Delburgo S, Thomsen B C, et al. Optimization of optical data transmitters for 40-Gb/s lightwave systems using frequency resolved optical gating. IEEE Photonics Technology Letters, 2002, 14(7): 971～974.

[35] Mollenauer L F, Stolen R H, Gordon J P. Experimental observation of picosecond pulse narrowing and solitons in optical fiber. Phys. Rev. Lett., 1980, 45(13): 1095～1098.

[36] Mollenauer L F. Nonlinear optics in fibers. Science, 2003, 3021: 996～997.

[37] Agrawal G P. Fiber optic communication systems. 3rd edition. New York: John Wiley & Sons, Inc., 2002.

[38] Agrawal G P. Nonlinear Fiber Optics. 5th edition. Uk: Academic Press, 2013.

[39] Ablowitz M J, Kaup D J, Newell A C, et al. The inverse scattering transform-fourier analysis for nonlinear problems. Studies in Applied Mathematics, 1974, 53(4): 249～315.

[40] Liu S L, Wang W Z, Xu J Z. Exact N-soliton solutions of the modified nonlinear Schrödinger equation. Physical Review E, 1993, 48 (4):3054～3059; Exact N-soliton solutions of the extended nonlinear Schrödinger equation. Physical Review E, 1994, 49 (6):5726～5730.

[41] Assefa S, Vlasov Y A. High-order dispersion in photonic crystal waveguides. Opt. Express, 2007,

15: 17562～17569.

[42] Mitschke F M, Mollenauer L F. Discovery of the soliton self-frequency shift. Opt. Lett., 1986, 11: 659～661.

[43] de Oliveira J R, de Moura M A, Hickmann J M, et al. Self-steepening of optical pulses in dispersive media. J. Opt. Soc. Am. B, 1992, 9: 2025～2027.

[44] Panoiu N C, Liu X P, Jr Osgood R M. Self-steepening of ultrashort pulses in silicon photonic nanowires. Opt. Lett., 2009, 34: 947～949.

[45] Skryabin D V, Luan F, Knight J C, et al. Soliton self-frequency shift cancellation in photonic crystal fibers. Science, 2003, 301: 1705～1708.

[46] Mitschke F M, Mollenauer L F. Experimental observation of interaction forces between solitons in optical fibers. Opt. Lett., 1987, 12: 355～357.

[47] Smith K, Mollenauer L F. Experimental observation of soliton interaction over long fiber paths: Discovery of a long-range interaction. Opt. Lett., 1989, 14: 1284～1286.

[48] Pinto A N, Agrawal G P, Ferreira da Rocha J. Effect of soliton interaction on timing jitter in communication systems. J. Lightwave Technol., 1998, 16: 515～519.

[49] Agrawal G P. Effect of intrapulse stimulated Raman scattering on soliton-effect pulse compression in optical fibers. Opt. Lett., 1990, 15: 224～226.

[50] Mamyshev P V, Mollenauer L F. Stability of soliton propagation with sliding-frequency guiding filters. Opt. Lett., 1994, 19: 2083～2085.

[51] Mollenauer L F, Stolen R H, Gordon J P, et al. Extreme picosecond pulse narrowing by means of soliton effect in single-mode optical fibers. Opt. Lett., 1983,8: 289～291.

[52] Ouzounov D, Hensley C, Gaeta A, et al. Soliton pulse compression in photonic band-gap fibers. Opt. Express, 2005,13: 6153～6159 .

[53] Mollenauer L F, Stolen R H. The soliton laser. Opt. Lett., 1984, 9: 13～15.

[54] Gredeskul S A, Kivshar Y S. Dark-soliton generation in optical fibers. Opt. Lett.， 1989， 14： 1281～1283.

[55] Litchinitser N, Agrawal G, Eggleton B, et al. High-repetition-rate soliton-train generation using fiber Bragg gratings. Opt. Express, 1998, 3: 411～417 .

[56] Gong Y, Shum P, Tang D, et al. 660GHz solitons source based on modulation instability in short cavity. Opt. Express, 2003, 11: 2480～2485.

[57] Zhang J D, Lin Q, Piredda G, et al. Optical solitons in a silicon waveguide. Opt. Express, 2007, 15: 7682～7688.

[58] Liu S L, Zheng H J. Experimental research on evolution of optical pulses into solitons in standard single mode fiber. Acta Opt. Sin, 2006, 26(9):1313～1318 (in Chinese).

[59] Chen C J, Wai P K A, Menyuk C R. Soliton switch using birefringent optical fibers. Opt. Lett., 1990, 15: 477～479.

[60] Wilson J, Stegeman G I, Wright E M. Soliton switching in an erbium-doped nonlinear fiber coupler. Opt. Lett., 1991, 16: 1653～1655.

[61] Locati F S, Romagnoli M, Tajani A, et al. Adiabatic femtosecond soliton active nonlinear directional coupler. Opt. Lett., 1992, 17: 1213～1215.

[62] Mollenauer L F, Smith K. Demonstration of soliton transmission over more than 4000 km in fiber with loss periodically compensated by Raman gain. Opt. Lett., 1988, 13: 675～677.

[63] Mollenauer L F, Neubelt M J, Evangelides S G, et al. Experimental study of soliton transmission over more than 10,000 km in dispersion-shifted fiber. Opt. Lett., 1990, 15: 1203～1205.

[64] Serkin VN, Hasegawa A. Novel Soliton Solutions of the nonlinear Schrödinger equation model. Phys. Rev. Lett, 2000, 85: 4502～4505; Soliton management in the nonlinear Schrödinger equation model with varying dispersion, nonlinearity, and gain. JETP Letters, 2000, 72 (2): 89～92.

[65] Kruglov V I, Peacock A C, Harvey J D. Exact self-similar solutions of the generalized nonlinear Schrödinger equation with distributed coefficients. Phys. Rev. Lett., 2003, 90(11): 113902.

[66] Hao R Y, Li L, Li Z H, et al. A new approach to exact soliton solutions and soliton interaction for the nonlinear Schrödinger equation with variable coefficients. Optics Communications, 2004, 236 (1-3): 79～86.

[67] Yang R C, Hao R Y, Li L, et al. Exact gray multi-soliton solutions for nonlinear Schrödinger equation with variable coefficients. Optics Communications, 2007, 253 (1-3): 177～185.

[68] Yang G Y, Hao R Y, Li L, et al. Cascade compression induced by nonlinear barriers in propagation of optical solitons. Optics Communications, 2006, 260 (1): 282～287.

[69] Zhang J L, Li B A, Wang M L. The exact solutions and the relevant constraint conditions for two nonlinear Schrödinger equations with variable coefficients. Chaos, Solitons & Fractals, 2009, 39 (2): 858～865.

[70] Wang J F, Li L, Jia S T. Exact chirped gray soliton solutions of the nonlinear Schrödinger equation with variable coefficients. Optics Communications, 2007, 274 (1): 223～230.

[71] Stolen R H, Ashikin A. Optical Kerr effect in glass waveguide. Appl. Phys. Lett., 1973, 22(6): 294～296.

[72] Kitayama K , Kimura Y, Okamoto K, et al. Optical sampling using an all-fiber optical Kerr shutter. Appl. Phys. Lett., 1985, 46(7): 623～625.

[73] Islam M N, Soccolich C E, Chen C J, et al. All-optical inverter with one picojoule switching energy. Electronics Letters, 1991, 27(2): 130～132.

[74] Asobe M, Kanamori T, Kubodera K. Ultrafast all-optical switching using highly nonlinear chalcogenide glass fiber. IEEE Photon. Technol. Lett., 1992, 4(4): 362～365.

[75] Islam M N, Sauer J R. GEO modules as a natural basis for all-optical fiber logic systems. IEEE J. Quantum Electron., 1991, 27(3): 843～848.

[76] Menyuk C R. Nonlinear pulse propagation in birefringent optical fibers. IEEE J. of Quantum Electron., 1987, 23(2): 174～176.

[77] Menyuk C R. Stability of solitons in birefringent optical fibers. II. Arbitrary amplitudes. J. Opt. Soc. Am. B, 1988, 5(2): 392～402.

[78] Menyuk C R. Stability of solitons in birefringent optical fibers. I: Equal propagation amplitudes. Optics Letters, 1987, 12(8): 614～616.

[79] Cao X D, Meyerhofer D D. Soliton collisions in optical birefringent fibers. J. Opt. Soc. Am. B, 1994, 11(2): 380～385.

[80] 唐雄燕, 叶培大. 双折射光纤中正交极化孤子碰撞的数值研究. 北京邮电大学学报, 1994,

17(2): 10～17.

[81] 江辉, 庞勇, 蒋佩璇. 双折射光纤中孤子碰撞的数值研究. 北京邮电大学学报, 1996, 19(4): 42～47.

[82] 黄洪涛, 聂再清. 线双折射光纤与正交极化孤子碰撞的研究. 中国激光, 1999, 26(6): 163～170.

[83] Manakov S V. On the theory of two-dimensional stationary self-focusing of electromagnetic waves. Sov. Phys. JETP, 1974, 38(2): 248～253.

[84] Smyth N F, Kath W L. Radiative losses due to pulse interactions in birefringent nonlinear optical fibers. Phys. Rev. E, 2001, 63(3): 036614.

[85] Kanna T, Lakshmanan M. Exact soliton solutions of coupled nonlinear Schrödinger equations: Shape-changing collisions, logic gates, and partially coherent solitons. Phys. Rev. E, 2003, 67(4): 046617～046625.

[86] Soljačić M, Steiglitz K, Sears S M, et al. Collisions of two solitons in an arbitrary number of coupled nonlinear Schrödinger equations. Phys. Rev. Lett., 2003, 90(25): 254102.

[87] Horikis T P, Elgin J N. Nonlinear optics in a birefringent optical fiber. Phys. Rev. E, 2004, 69(1): 016603.

[88] Zheng H J, Liu S L, Wu C Q, et al. Experimental study on pulse propagation characteristics at normal dispersion region in dispersion flatted fibers. Optics & Laser Technology, 2012, 44(4): 763～766.

[89] Zheng H J, Liu S L, Li X, et al. Pulse compression in dispersion-decreasing-like fibers. Optics & Laser Technology, 2011, 43(7): 1321～1324.

[90] Zheng H J, Wu C Q, Wang Z, et al. Propagation characteristics of chirped soliton in periodic distributed amplification systems with variable coefficients. Optik, 2012, 123(9): 818～822.

[91] Zheng H J, Liu S L, Li X, et al. Propagation stability of a chirped soliton in birefringent fibers. Journal of Physics:Conference Series, 2011, 276(1): 012040-1-6.

[92] Zheng H J, Liu S L, Xu J P. Effect of initial linear chirp on collision characteristics of two solitons in the birefringent fiber. Chinese Physics, 2007, 16(7)：2023～2027.

第 6 章　啁啾脉冲的超连续谱

6.1　啁啾脉冲在凸形色散平坦光纤中的超连续谱

近年来，光纤超连续谱广泛应用于波分复用光通信系统及其关键光电子器件、超短脉冲产生、光学相干层析和光谱分析等重要领域，其研究得到了广泛关注[1-23]。人们对在色散位移光纤、色散平坦光纤、色散平坦渐减光纤、光子晶体光纤等多种光纤中产生的超连续谱进行了理论和实验研究，其中凸形色散分布平坦光纤或平坦渐减光纤中产生的超连续谱特性较好[24-30]。文献[30]以无啁啾的双曲正割脉冲为入射脉冲，在凸形色散分布平坦光纤反常色散区获得了宽带、平坦的超连续谱。然而，实际激光脉冲源通常为具有较大频率啁啾的高斯脉冲且频率啁啾可以通过改变传输光纤长度、采用啁啾光栅技术或预啁啾技术等进行调节[1]，本书以啁啾高斯脉冲为入射脉冲，在凸形和凹形色散分布平坦光纤反常色散区，研究了线性频率啁啾对高斯脉冲产生超连续谱的影响，并与啁啾双曲正割脉冲产生超连续谱的情况进行了比较，为我们利用啁啾脉冲获得最佳的超连续谱和实现波分复用光纤通信系统超连续谱光源的优化设计提供了重要依据。

6.1.1　凸形色散平坦光纤中产生超连续谱的理论模型

脉冲产生超连续谱满足传输方程(1-21)，为了方便利用方程(1-21)研究脉冲的传输特性，引入

$$\xi = z/L_D, \quad \tau = \frac{T}{T_0} = \frac{t-\beta_1 z}{T_0}, \quad A(z,\tau) = \sqrt{P_0}\,U(z,\tau)$$

$$L_D = \frac{T_0^2}{|\beta_2|}, \quad L_{\mathrm{NL}} = \frac{1}{\gamma P_0}, \quad N^2 = \frac{L_D}{L_{\mathrm{NL}}} = \frac{\gamma P_0 T_0^2}{|\beta_2|}, \quad u = NU$$

因本节讨论的最高阶色散为 6 阶，将式(1-21)修正后的归一化方程为

$$\frac{\partial u}{\partial \xi} = \mathrm{i}\sum_{n=2}^{\infty}\frac{\mathrm{i}^n \beta_n}{n!|\beta_2|T_0^{n-2}}\frac{\partial^n u}{\partial \tau^n} + \mathrm{i}|u|^2 u - s\frac{\partial(|u|^2 u)}{\partial \tau} - \mathrm{i}\,\tau_{\mathrm{R}} u\frac{\partial |u|^2}{\partial \tau} - \frac{1}{2}\Gamma u, \quad n = 2,\cdots,6 \tag{6-1}$$

式中，T_0 是脉冲的半宽度，$s = \dfrac{1}{\omega_0 T_0}$，$\tau_{\mathrm{R}} = \dfrac{T_{\mathrm{R}}}{T_0}$，$\Gamma = \alpha L_D = \alpha T_0^2/|\beta_2|$。

光纤色散参量 D 与群速度色散系数以及脉冲中心波长的关系为[1]

$$D=\frac{\mathrm{d}\beta_1}{\mathrm{d}\lambda},\quad \omega=\frac{2\pi c}{\lambda},\quad \beta_m=\left(\frac{\mathrm{d}^m\beta}{\mathrm{d}\omega^m}\right)_{\omega=\omega_0},\quad m=0,1,2,\cdots \tag{6-2}$$

实际应用中只能测量得到光纤色散参量 D，而不能直接测得群速度色散系数 β_n，因此，作者由式(6-2)求得光纤中脉冲中心波长 λ_0 处的各阶(本节讨论的最高阶为 6 阶)群速度色散系数的具体表达式为

$$\beta_2(\lambda_0)=\left[\left(-\frac{\lambda^2}{2\pi c}\right)\cdot D\right]_{\lambda=\lambda_0} \tag{6-3a}$$

$$\beta_3(\lambda_0)=\left[\left(-\frac{\lambda^2}{2\pi c}\right)^2\left(\frac{\mathrm{d}D}{\mathrm{d}\lambda}+\frac{2D}{\lambda}\right)\right]_{\lambda=\lambda_0} \tag{6-3b}$$

$$\beta_4(\lambda_0)=\left[\left(-\frac{\lambda^2}{2\pi c}\right)^3\left(\frac{\mathrm{d}^2D}{\mathrm{d}\lambda^2}+\frac{6}{\lambda}\frac{\mathrm{d}D}{\mathrm{d}\lambda}+\frac{6D}{\lambda^2}\right)\right]_{\lambda=\lambda_0} \tag{6-3c}$$

$$\beta_5(\lambda_0)=\left[\left(-\frac{\lambda^2}{2\pi c}\right)^4\left(\frac{\mathrm{d}^3D}{\mathrm{d}\lambda^3}+\frac{12}{\lambda}\frac{\mathrm{d}^2D}{\mathrm{d}\lambda^2}+\frac{36}{\lambda^2}\frac{\mathrm{d}D}{\mathrm{d}\lambda}+\frac{24D}{\lambda^3}\right)\right]_{\lambda=\lambda_0} \tag{6-3d}$$

$$\beta_6(\lambda_0)=\left[\left(-\frac{\lambda^2}{2\pi c}\right)^5\left(\frac{\mathrm{d}^4D}{\mathrm{d}\lambda^4}+\frac{20}{\lambda}\frac{\mathrm{d}^3D}{\mathrm{d}\lambda^3}+\frac{120}{\lambda^2}\frac{\mathrm{d}^2D}{\mathrm{d}\lambda^2}+\frac{240}{\lambda^3}\frac{\mathrm{d}D}{\mathrm{d}\lambda}+\frac{120D}{\lambda^4}\right)\right]_{\lambda=\lambda_0} \tag{6-3e}$$

凸形色散分布平坦光纤的色散参量与脉冲中心波长满足[30]

$$D(\lambda)=D_0+\frac{k}{2}(\lambda-\lambda_D)^2 \tag{6-4}$$

式中，D_0 为光纤的色散峰值，k 为光纤色散参量 D 关于波长的二阶微分常量，λ_D 为色散峰值波长，光纤损耗为 0.2dB/km，由式(6-4)可得

$$\frac{\mathrm{d}D}{\mathrm{d}\lambda}=k(\lambda-\lambda_D),\quad \frac{\mathrm{d}^2D}{\mathrm{d}\lambda^2}=k,\quad \frac{\mathrm{d}^3D}{\mathrm{d}\lambda^3}=0,\quad \frac{\mathrm{d}^4D}{\mathrm{d}\lambda^4}=0 \tag{6-5}$$

本节中，令 $\lambda_D=\lambda_0$，选择入射脉冲的中心波长 $\lambda_0=1550\text{nm}$，$D_0=0.2\text{ps/(nm}\cdot\text{km)}$，$k=-0.0003\text{ps/(nm}^3\cdot\text{km)}$。将上述参量、式(6-4)和式(6-5)代入式(6-3)，可以得到式(6-1)中各阶群速度色散系数的具体数值。然后，分别以具有相同时域半峰全宽 $T_{\text{FWHM}}=0.2\text{ps}$ 和脉冲幅值 $N=3$ 的线性啁啾高斯脉冲(C 是线性啁啾参量)

$$u(0,\tau)=N\exp\left(-\frac{\tau^2}{2}\right)\exp\left(-\frac{\mathrm{i}C\tau^2}{2}\right) \tag{6-6}$$

和线性啁啾双曲正割脉冲

$$u(0,\tau)=N\operatorname{sech}(\tau)\exp\left(-\frac{\mathrm{i}C\tau^2}{2}\right) \tag{6-7}$$

作为入射脉冲，采用分步傅里叶方法，按照归一化方程(6-1)数值研究初始啁啾对在凸形色散分布平坦光纤反常色散区产生超连续谱的影响。

6.1.2 凸形色散平坦光纤中的超连续谱数值计算与分析

6.1.2.1 脉冲的频谱质心随啁啾参量 C 的变化

频谱质心可以描述脉冲频谱向长、短波长方向移动的情况，定义为

$$\omega_{\text{cent}}(\xi)=\frac{\int_{-\infty}^{+\infty}\omega\left|u(\xi,\omega)\right|^2\mathrm{d}\omega}{\int_{-\infty}^{+\infty}\left|u(\xi,\omega)\right|^2\mathrm{d}\omega} \tag{6-8}$$

图 6-1 所示是脉冲的频谱质心随啁啾参量 C 的变化。横坐标是传输距离，归一化到孤子周期(soliton periods) z_0；纵坐标是脉冲频谱质心。实线所示对应高斯脉冲的入射情况，虚线所示对应双曲正割脉冲的情况。曲线 1 和 4、2 和 5、3 和 6 分别对应 $C=-1$、0 和 1。

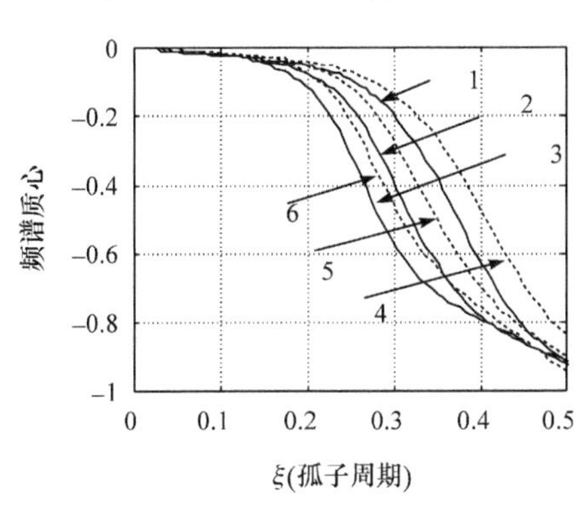

图 6-1　高斯(实线)和双曲正割(虚线)脉冲的频谱质心

曲线 1 和 4、2 和 5、3 和 6 分别对应 $C=-1$、0 和 1

当 $0<\xi\leqslant 0.32z_0$ 时，高斯脉冲频谱质心随传输距离和啁啾参量 C 的增加而红移，红移速度逐渐增加，高斯脉冲频谱质心比双曲正割脉冲的红移大且红移速度增加。当 $0.32z_0<\xi\leqslant 0.5z_0$ 时，随传输距离和啁啾参量 C 的增加，高斯脉冲频谱质心的红移及其速度逐渐减小，不同 C 对应的频谱质心红移的差距随传输距离增加而减小，$C<0$ 对应的频谱质心红移及其速度仍然大于双曲正割脉冲的，$C\geqslant 0$ 对应的频谱质心红移及其速度与双曲正割脉冲的差距随距离增大而减小。脉冲频谱质心红移是脉冲拉曼自频移导致脉冲频谱中心向长波长漂移的结果，正啁啾对拉曼效应起增强作用，负啁啾对拉曼效应起减弱作用。因为当非线性项仅考虑自相位调制时，脉冲的频谱质心微蓝移，且随啁啾参量 C 增加而增大；仅考虑自相位调制和自变陡效应时，脉冲的频谱质心蓝移稍大，且随啁啾参量 C 增加而继续增大。

6.1.2.2　脉冲的方均根脉宽随啁啾参量 C 的变化

在光纤中形成超连续谱的过程中，脉冲的时域波形和频谱变化复杂，脉冲时域和频谱半峰全宽不再是脉冲特性的真实量度，通常采用脉冲的方均根脉宽和方均根谱宽来描述脉冲时域和频谱的变化规律。脉冲的方均根脉宽定义为

$$W_{\tau}=\left[\frac{\int_{-\infty}^{+\infty}\tau^{2}\left|u(\xi,\tau)\right|^{2}\mathrm{d}\tau}{\int_{-\infty}^{+\infty}\left|u(\xi,\tau)\right|^{2}\mathrm{d}\tau}-\left(\frac{\int_{-\infty}^{+\infty}\tau\left|u(\xi,\tau)\right|^{2}\mathrm{d}\tau}{\int_{-\infty}^{+\infty}\left|u(\xi,\tau)\right|^{2}\mathrm{d}\tau}\right)^{2}\right]^{1/2} \tag{6-9}$$

图 6-2 所示是脉冲的方均根脉宽随啁啾参量 C 的变化。纵坐标是脉冲方均根脉宽，横坐标是归一化传输距离。

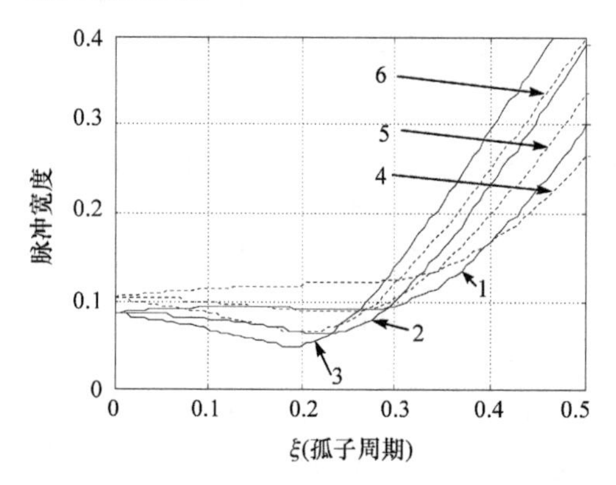

图 6-2　高斯(实线)和双曲正割(虚线)脉冲的方均根脉宽

曲线 1 和 4、2 和 5、3 和 6 分别对应 $C=-1$、0 和 1

由图 6-2 可见，在$-1.5\leqslant C<0$ 的情况下，当 $0<\xi\leqslant 0.3z_0$ 时，高斯脉冲随传输距离增加先稍展宽，后稍压缩；啁啾越大，脉冲压缩越小(曲线 1 所示)；这表明初始负啁啾减弱了非线性效应。当 $0.3z_0<\xi\leqslant 0.5z_0$ 时，高斯脉冲随传输距离增加迅速单调展宽；$|C|$越大，展宽越慢；脉冲的这种展宽主要由强非线性导致的明显边峰引起，与仅由色散引起的脉冲展宽不同。在 $C\geqslant 0$ 的情况下，当 $0<\xi\leqslant 0.3z_0$ 时，高斯脉冲随传输距离增加先压缩后展宽；C 越大，脉冲压缩越大(曲线 2 和 3 所示)；这表明初始正啁啾增强了非线性效应。当 $0.3z_0<\xi\leqslant 0.5z_0$ 时，高斯脉冲随传输距离增加迅速单调展宽；$|C|$越大，展宽越快。

在 $\xi>0.3z_0$ 后，正啁啾时对应的脉冲展宽比负啁啾时要快且宽，这表明正啁啾对脉冲展宽的影响比负啁啾大；高斯脉冲比双曲正割脉冲展宽得快且宽。

6.1.2.3　脉冲的方均根频谱宽度随啁啾参量 C 的变化

脉冲的方均根谱宽定义为

$$W_{\omega}=\left[\frac{\int_{-\infty}^{+\infty}\omega^{2}\left|u(\xi,\omega)\right|^{2}\mathrm{d}\omega}{\int_{-\infty}^{+\infty}\left|u(\xi,\omega)\right|^{2}\mathrm{d}\omega}-\left(\frac{\int_{-\infty}^{+\infty}\omega\left|u(\xi,\omega)\right|^{2}\mathrm{d}\omega}{\int_{-\infty}^{+\infty}\left|u(\xi,\omega)\right|^{2}\mathrm{d}\omega}\right)^{2}\right]^{1/2} \tag{6-10}$$

图 6-3 所示是脉冲的方均根谱宽随啁啾参量 C 的变化。纵坐标是脉冲方均根谱宽，横坐标是归一化传输距离。

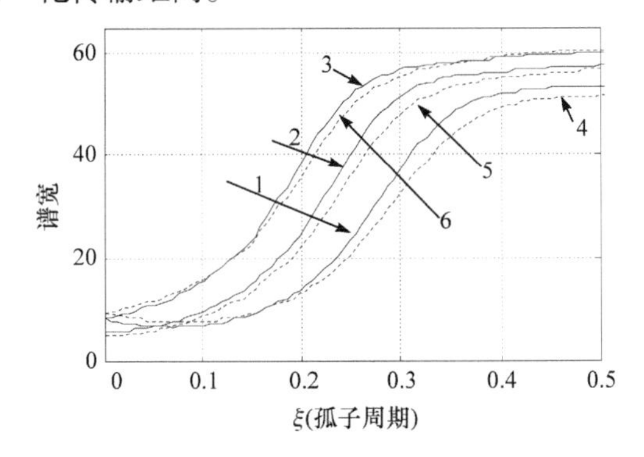

图 6-3　高斯(实线)和双曲正割(虚线)脉冲的方均根频谱宽度

曲线 1 和 4、2 和 5、3 和 6 分别对应 C=−1、0 和 1

当 $0<\xi\leqslant 0.32z_0$ 时，高斯脉冲频谱随传输距离和啁啾参量 C 的增加逐渐展宽，展宽速度逐渐加快；高斯脉冲频谱比双曲正割脉冲的稍宽且展宽得快。当 $0.32z_0<\xi\leqslant 0.5z_0$ 时，两脉冲谱宽随传输距离增加几乎不变，两脉冲谱宽随啁啾参量 C 增加而稍微增加，两者差距很小。在 $0.32z_0$ 附近，$C\geqslant 0$ 对应的频谱已基本展宽到最大值，$C<0$ 对应的频谱仍有较大的展宽余地。由此可得，若想得到较宽的超连续谱，正啁啾脉冲入射所需传输光纤长度小于负啁啾脉冲情况。

6.1.2.4　脉冲的超连续谱及其随啁啾参量 C 的变化

高峰值功率(N=3)高斯脉冲在光纤中形成超连续谱时，首先入射脉冲频谱整体展宽、频谱质心微红移、脉冲压缩、峰值功率增加；在随后传输中，频谱中心频率强度明显减弱，其两侧频谱强度迅速增加，脉冲继续压缩，主脉冲边缘附近出现明显的边锋，形成高阶孤子；进一步传输，在高阶色散和非线性共同作用下，在脉冲第一次展宽附近形成较平坦的超连续谱，随后，超连续谱的宽度随传输距离变化不大，其平坦度随传输距离的增加而变化。其中，三阶

(β_3)、五阶(β_5)和六阶色散(β_6)对超连续谱影响很小，可忽略；二阶(β_2)、四阶色散(β_4)和自相位调制(SPM)对超连续谱的形成起关键作用；初始啁啾对超连续谱特性影响较大；自变陡效应和拉曼散射效应对超连续谱影响较小。图 6-4 所示是 C=0 时脉冲在凸形色散平坦光纤中的超连续谱。实线所示对应高斯脉冲在 $0.3z_0$ 处的超连续谱，点线所示对应双曲正割脉冲在 $0.32z_0$ 处的情况。横坐标是波长，单位为纳米(nm)；纵坐标是频谱强度，归一到入射脉冲频谱峰值，单位为分贝(dB)。图 6-4～图 6-6 的坐标相同，且均考虑了最佳超连续谱与光纤长度的关系。

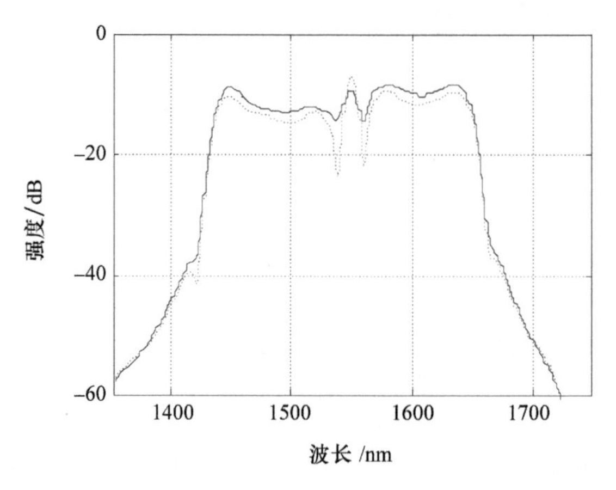

图 6-4　C=0 时高斯(实线，$0.3z_0$)和双曲正割脉冲(点线，$0.32z_0$)的超连续谱

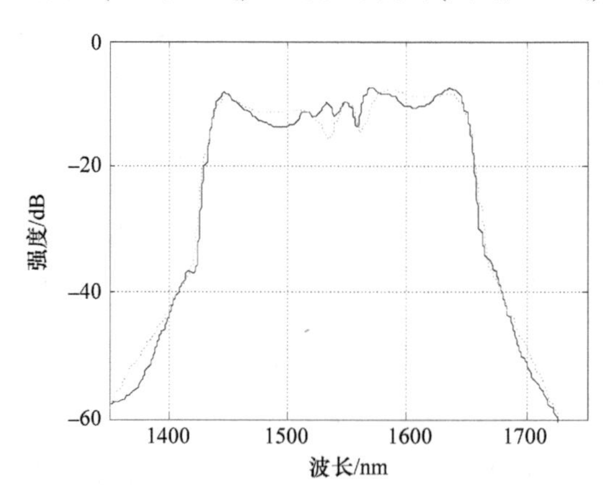

图 6-5　C=−0.5(实线，$0.33z_0$)和 0.5(点线，$0.27z_0$)时高斯脉冲的超连续谱

由图 6-4 可见，双曲正割脉冲的超连续谱与文献[20]一致。高斯脉冲的超连续谱达到 200nm 以上，比双曲正割脉冲的更平坦且稍宽。在泵浦脉冲中心波长

附近，高斯脉冲超连续谱仍有抽运残余，但比双曲正割脉冲的明显减小。在频谱边缘附近，高斯脉冲超连续谱关于中心频谱基本对称，比双曲正割脉冲的稍宽，频谱强度略大；高斯脉冲超连续谱边缘下降比双曲正割脉冲的要快。高斯脉冲超连续谱的整体特性优于双曲正割脉冲，这与高斯入射脉冲频谱比具有相同半峰全宽的双曲正割脉冲频谱宽得多有关。

图 6-5 所示是啁啾参量 C 不同时高斯脉冲的超连续谱。实线所示对应 $C=-0.5$ 时在 $0.33z_0$ 处的情况，点线所示对应 $C=0.5$ 时在 $0.27z_0$ 处情况。由图 6-5 可见，啁啾参量 C 越大，获得最佳超连续谱所需的光纤长度越短。$|C|<1$ 情况下，在超连续谱中心频率两侧和超连续谱边缘，负啁啾使频谱强度增强，正啁啾使之减弱。负啁啾时的超连续谱特性比正啁啾时的特性稍差，两种情况下的超连续谱特性均稍劣于无啁啾时的超连续谱特性。

$|C|\geqslant 1$ 情况下，负啁啾的超连续谱平坦度明显劣于正啁啾情况，两种情况下的超连续谱特性均明显劣于无啁啾时的超连续谱特性，如图 6-6 所示。

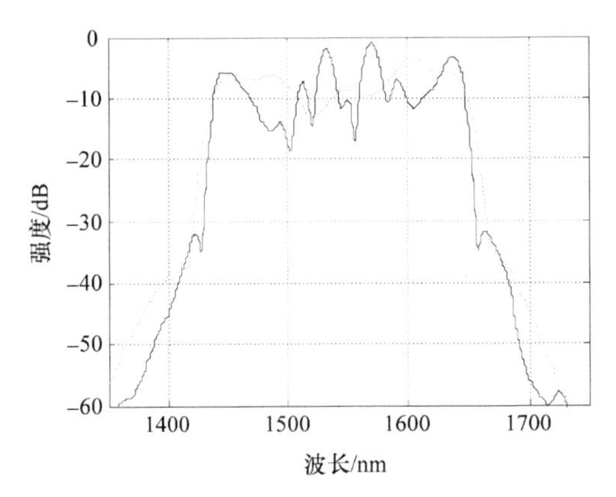

图 6-6　$C=-2$(实线，$0.46z_0$)和 2(点线，$0.22z_0$)时高斯脉冲的超连续谱

6.2　啁啾脉冲在凹形色散平坦光纤中的超连续谱

6.2.1　凹形色散平坦光纤中产生超连续谱的理论模型

皮秒、亚皮秒量级光脉冲在光纤中传输满足修正后的归一化方程(6-1)，光纤中光脉冲中心波长λ_0处的各阶(讨论的最高阶为 6 阶)群速度色散系数的具体表达式满足公式(6-3)。

凹形色散平坦光纤的色散参量 D 与脉冲中心波长λ满足

$$D(\lambda)=D_0+k_1(\lambda-\lambda_D)^2+k_2(\lambda-\lambda_D)^4 \tag{6-11}$$

式中，D_0为光纤的色散峰值，k_1和k_2为光纤色散参量D关于波长的色散系数，λ_D为$D(\lambda)=D_0$时的波长，光纤损耗为0.2dB/km。

本节中，令$\lambda_D=\lambda_0$，选择入射光脉冲的中心波长λ_0=1550nm，D_0=0.2ps/(nm · km)，k_1=−0.000115ps/(nm^3 · km)，k_2=8.5 × 10^{-9}ps/(nm^5 · km)。将上述参量、式(6-11)及其关于波长的多次导数代入式(6-3)，可以得到式(6-1)中各阶(本章讨论的最高阶为 6 阶)群速度色散系数的具体数值。然后，以时域半峰全宽T_{FWHM}=0.2ps 和脉冲幅值N=3 的线性啁啾高斯光脉冲(C是线性啁啾参量)

$$U(0,\tau)=N\exp\left(-\frac{\tau^2}{2}\right)\exp\left(-\frac{\mathrm{i}C\tau^2}{2}\right) \tag{6-12}$$

作为入射光脉冲，采用分步傅里叶方法，按照归一化方程(6-1)数值研究了在凹形色散平坦光纤中产生超连续谱的特性。数值计算时，光脉冲时域窗口为(−80，80)，采样点数为 1024。

6.2.2 凹形色散平坦光纤中的超连续谱数值计算与分析

6.2.2.1 光脉冲的频谱质心随啁啾参量C的变化

频谱质心可以描述光脉冲频谱向长、短波长方向移动的情况，定义为

$$\omega_{\mathrm{cent}}(\xi)=\frac{\int_{-\infty}^{+\infty}\omega\left|u(\xi,\omega)\right|^2\mathrm{d}\omega}{\int_{-\infty}^{+\infty}\left|u(\xi,\omega)\right|^2\mathrm{d}\omega} \tag{6-13}$$

图 6-7 所示是高斯光脉冲的频谱质心随传输距离和啁啾参量C的变化。横坐标是归一化传输距离，单位为孤子周期z_0；纵坐标是脉冲频谱质心。在无啁啾(C=0)情况下，当$0<\xi<0.255z_0$时，脉冲频谱质心随传输距离的增加而红移，且红移速度逐渐增加；在$\xi=0.255z_0$附近，脉冲频谱质心出现轻微振荡现象；当$\xi>0.255z_0$时，脉冲频谱质心随传输距离的增加而红移，且红移速度迅速增加。存在啁啾的情况下，脉冲频谱质心出现振荡的距离随啁啾参量C的增加而减小；出现振荡之后，脉冲频谱质心随啁啾参量C的增加而红移且红移速度迅速增加。研究中单独考虑了自相位调制和二阶色散，然后逐步考虑自变陡效应、脉冲内拉曼效应和其他高阶色散，结果表明光脉冲频谱质心红移主要是由脉冲内拉曼自频移所致，正啁啾起增强作用，负啁啾起减弱作用。在高阶色散(高阶色散具有重要作用)与非线性效应作用下光脉冲主峰与边峰进行能量的重新分布，脉冲频谱变化剧烈，出现了频谱质心振荡现象。

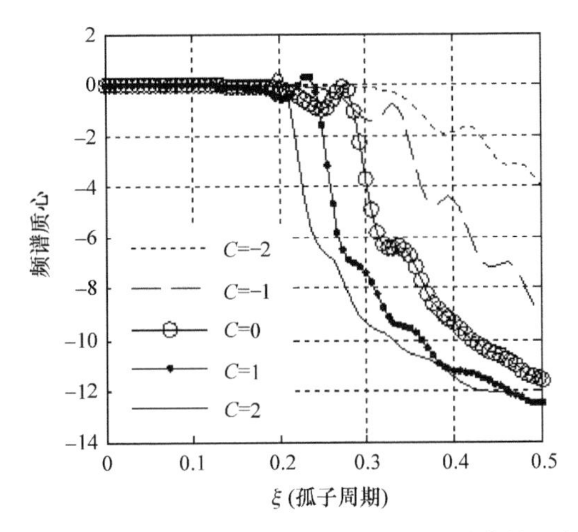

图 6-7　高斯脉冲的频谱质心随传输距离和啁啾参量 C 的变化

6.2.2.2　光脉冲的方均根频谱宽度随啁啾参量 C 的变化

在光纤中形成超连续谱的过程中，光脉冲的时域波形和频谱变化复杂，脉冲频谱半峰全宽不再是脉冲特性的真实量度，通常采用脉冲的方均根谱宽来描述脉冲时域和频谱的变化规律。光脉冲的方均根谱宽定义为

$$W_{\omega}=\left[\frac{\int_{-\infty}^{+\infty}\omega^{2}\left|u(\xi,\omega)\right|^{2}\mathrm{d}\omega}{\int_{-\infty}^{+\infty}\left|u(\xi,\omega)\right|^{2}\mathrm{d}\omega}-\left(\frac{\int_{-\infty}^{+\infty}\omega\left|u(\xi,\omega)\right|^{2}\mathrm{d}\omega}{\int_{-\infty}^{+\infty}\left|u(\xi,\omega)\right|^{2}\mathrm{d}\omega}\right)^{2}\right]^{1/2} \tag{6-14}$$

图 6-8 所示是高斯光脉冲的方均根谱宽随传输距离和啁啾参量 C 的变化。纵坐标是脉冲方均根谱宽。无啁啾情况下，当 $0<\xi\leqslant 0.3z_0$ 时，脉冲频谱宽度随传输距离增加逐渐展宽，展宽速度逐渐加快，并逐渐展宽到最大值；当 $0.3z_0<\xi\leqslant 0.5z_0$ 时，在高阶色散(高阶色散起关键作用)与非线性效应作用下，脉冲频谱平坦度急剧劣化，导致脉冲方均根频谱随传输距离增加逐渐变窄。在传输相同距离的情况下，方均根频谱达到最大值之前，谱宽及其展宽速度随啁啾参量 C 值的增加而增加。在 $0.3z_0$ 附近，$C=0$ 对应的频谱已基本展宽到最大值，$C>0$ 对应的频谱已经开始变窄，$C<0$ 对应的频谱仍有较大的展宽余地。由此可得，若想得到较宽较好的超连续谱，正啁啾光脉冲入射所需传输光纤长度要小于负啁啾脉冲情况；其他参量变化时，也存在最佳超连续谱与光纤长度的变化问题。图 6-9～图 6-13 均考虑了产生最佳超连续谱与光纤长度的关系。

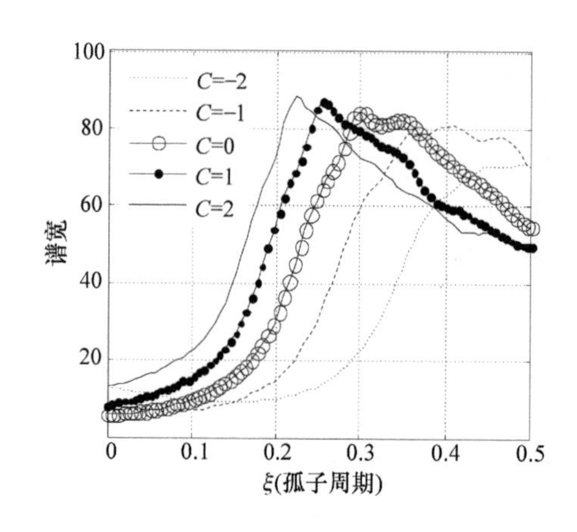

图 6-8　高斯脉冲方均根频谱宽度随距离和啁啾参量 C 的变化

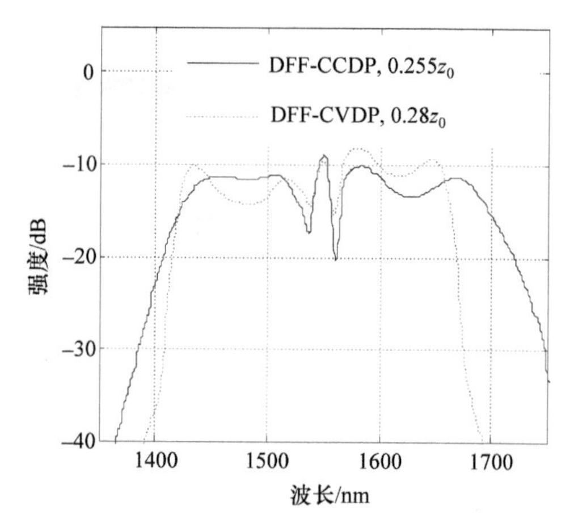

图 6-9　C=0 时高斯脉冲在凸形和凹形色散分布平坦光纤中的超连续谱

6.2.2.3　超连续谱及其随啁啾参量 C 的变化

高斯光脉冲在光纤中形成超连续谱时，入射光脉冲压缩、峰值功率增加、频谱中心频率强度逐渐减弱，其两侧频谱强度增加，频谱逐渐整体展宽，频谱质心微红移，主脉冲边缘附近出现明显的边峰；进一步传输，在高阶色散和非线性共同作用下，脉冲形成较平坦的超连续谱，随后，超连续谱的平坦度随传输距离的增加而变化。其中，三阶色散对超连续谱影响较小，可忽略；二阶和四阶色散对超连续谱的形成起关键作用；五阶色散对超连续谱平坦性影响较大；六阶色散对超连续谱宽度有较大影响。自变陡效应和拉曼散射效应对超连续谱影响较小。入射光脉冲峰值功率过小，超连续谱的整体特性很差；入射光脉冲峰值功率过大，自相位调制效应过强，虽然有利于脉冲的压缩和频谱的展宽，但是脉冲边峰越多越强，会导致频谱变化剧烈、超连续谱不平坦。入射光脉冲峰值功率的最佳值约为 N=3。现以 N=3 为例讨论。图 6-9 所示是 C=0 时高斯光脉冲在凹形色散分布平坦光纤(DFF-CCDP)和凸形色散分布平坦光纤(DFF-CVDP)中形成的超连续谱。横坐标是波长，单位为纳米(nm)；纵坐标是频谱强度，归一到入射脉冲初始频谱峰值，单位为分贝(dB)。

由图 6-9 可见，在−20dB 处，光脉冲在 DFF-CCDP 中产生的超连续谱达到 312.4nm，在 DFF-CVDP 中产生的超连续谱达到 252.7nm，相同情况下前者比后者宽得多，在−20dB 处增宽约 60nm。这主要是由光纤的色散分布类型不同导致的，DFF-CCDP 的色散参量 D 比 DFF-CVDP 的色散参量增加了波长的四次方项，导致了光纤的色散曲线呈明显的凹形分布，在较大的波长范围内，其色散值几乎不变，而 DFF-CVDP 的色散值变化稍大。将在 DFF-CCDP 和 DFF-CVDP 中产生的超连续谱进行比较，在短波区前者比后者更平坦，谱宽相差稍小；在长波区前者比后者更宽，两者平坦度相差不大。在泵浦脉冲中心波长附近，两者都存在抽运残余，前者比后者稍大。总之，前者的整体特性优于后者。

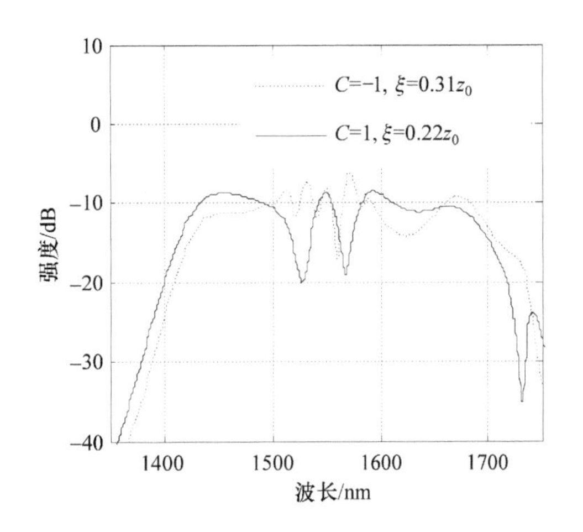

图 6-10 $C=-1$ 和 1 时高斯脉冲在 DFF-CCDP 中的最佳超连续谱

图 6-10 所示是啁啾参量 C 不同时高斯脉冲在 DFF-CCDP 中的最佳超连续谱。以下超连续谱的研究都是在 DFF-CCDP 中。由图 6-10 可见，在超连续谱中心频率两侧，负啁啾使频谱强度增强，正啁啾使之减弱。负啁啾时的最佳超连续谱整体平坦度优于正啁啾时的平坦度，两者的超连续谱宽度相差不大。负啁啾对应的频谱比正啁啾的频谱稍向长波长方向移动，这是因为前者对应的传输距离比后者对应距离长，使脉冲内拉曼效应稍明显。总之，$|C|\leqslant 1$ 情况下，负啁啾时的超连续谱特性优于正啁啾时的谱特性。

6.2.2.4 超连续谱随入射光脉冲宽度的变化

光脉冲在光纤中的传输特性与脉宽密切相关，那么光脉冲脉宽对在光纤中产生超连续谱的影响较大。图 6-11 所示是保持其他参数不变($N=3$，$C=0$，$k_1=-0.000115\text{ps}/(\text{nm}^3\cdot\text{km})$，$k_2=8.5\times10^{-9}\text{ps}/(\text{nm}^5\cdot\text{km})$)，光纤中超连续谱随入射光脉冲脉宽的变化。由图 6-11 可见，$T_{\text{FWHM}}=0.1\text{ps}$ 时，超连续谱在长波区平坦度最好，在短波区平坦度较差，在−20dB 处超连续谱宽度达到 364nm，需要传输距离 $\xi=0.3z_0=6.7\text{m}$；$T_{\text{FWHM}}=0.2\text{ps}$ 时，超连续谱在长波区平坦度稍差，在短波区平坦度最好，在−20dB 处超连续谱宽度达到 312.4nm，需要传输距离 $\xi=0.255z_0=22.7\text{m}$；$T_{\text{FWHM}}=0.3\text{ps}$ 时，超连续谱特性最差，在−20dB 处谱宽仅达到 209nm，需要传输距离 $\xi=0.24z_0=48\text{m}$。可见，入射光脉冲宽度越窄，产生的超连续谱越宽，获得最佳超连续谱所需光纤长度越短。$T_{\text{FWHM}}=0.2\text{ps}$ 时的超连续谱的整体特性最好。在超连续谱的应用中，可以根据实际情况，选择不同的入射光脉冲宽度。

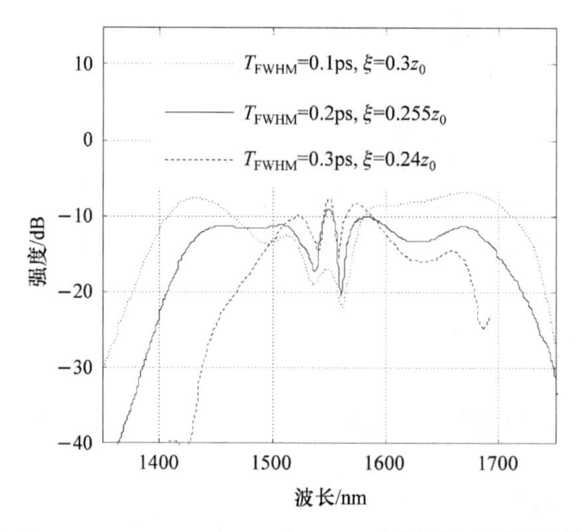

图 6-11 T_{FWHM}=0.1ps、0.2ps 和 0.3ps 时高斯脉冲的超连续谱

6.2.2.5 超连续谱随色散参量 D_0、k_1 和 k_2 的变化

光脉冲在光纤中的传输特性与光纤色散参量密切相关，那么光纤色散参量对在光纤中产生的超连续谱也有着较大影响。图 6-12 所示是保持其他参数不变，光纤中超连续谱随光纤色散参量 D_0 的变化。由图 6-12 可见，D_0=0.3ps/(nm · km)时最佳超连续谱需要传输距离ξ=0.245z_0=14.5m，超连续谱明显较宽，长波区的平坦度比短波区的好，长波区的频谱强度比短波区的大得多，且中心频率附近的频谱强度变化剧烈；D_0=0.1ps/(nm · km)时最佳超连续谱需要传输距离ξ=0.31z_0=55.1m，超连续谱虽然在长波区或短波区的平坦度均很好，但是宽度最窄、整体平坦度稍差；D_0=0.2ps/(nm · km)时最佳超连续谱需要传输距离ξ=0.255z_0=22.7m，超连续谱整体特性最好。超连续谱的宽度随 D_0 增大而增加，获得最佳超连续谱所需光纤长度随之而减小。

保持其他参数不变，光纤中超连续谱随光纤色散参量 k_1 和 k_2 的变化规律如图 6-13 所示。由图 6-13 可见，0.5k_1 和 0.5k_2 时，最佳超连续谱需要传输距离ξ=0.245z_0=21.8m，k_1 和 k_2 时最佳超连续谱需要传输距离ξ=0.255z_0=22.7m，2k_1 和 2k_2 时最佳超连续谱需要传输距离ξ=0.3z_0=26.7m。超连续谱的宽度随 k_1 和 k_2 增大而减小，获得最佳超连续谱所需光纤长度随之而增加。光纤色散参量 D_0、k_1 和 k_2 过小或过大，产生的超连续谱的整体特性较差。在优化光纤参量的设计中，应根据实际情况，综合考虑选择不同的色散参量。

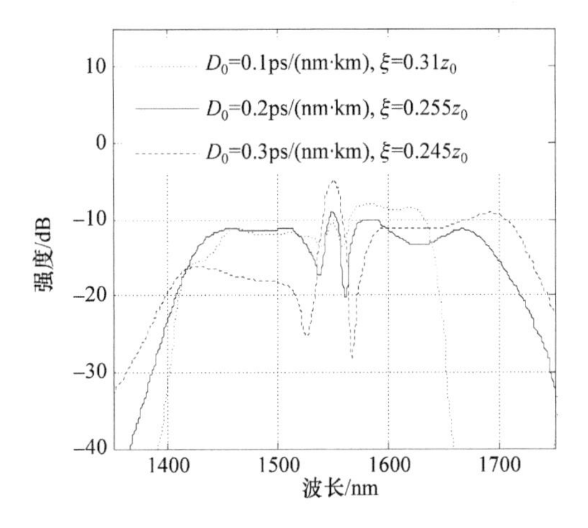

图 6-12　T_{FWHM}=0.2ps，D_0=0.1、0.2 和 0.3ps/(nm · km)时高斯脉冲的超连续谱

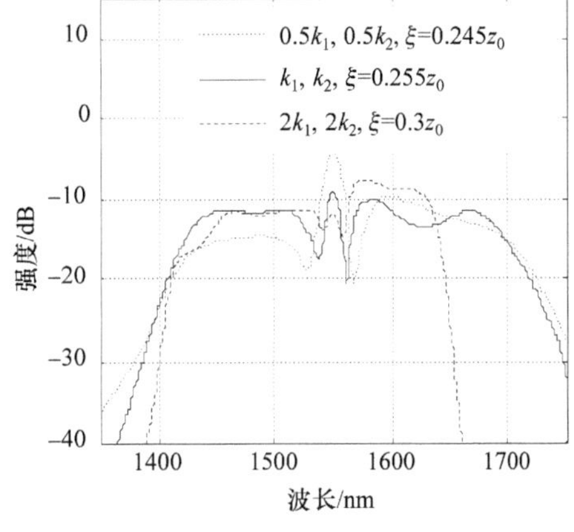

图 6-13　0.5k_1 和 0.5k_2、k_1 和 k_2、2k_1 和 2k_2 时高斯脉冲的超连续谱

6.3　本章小结

本章提出采用凸形色散平坦光纤和凹形色散平坦光纤产生超连续谱，利用分步傅里叶方法数值研究了高斯光脉冲在两种类型色散平坦光纤反常色散区产生的超连续谱；研究了线性频率啁啾对高斯脉冲在两种光纤中产生超连续谱的影响，并与啁啾双曲正割脉冲产生超连续谱的情况进行了比较[31, 32]。

结果表明，在凸形色散平坦光纤中，无啁啾高斯脉冲的超连续谱达到 200nm 以上，其特性优于双曲正割脉冲。获得最佳超连续谱所需光纤长度随啁啾参量 C 增加而缩短；负啁啾时的超连续谱特性劣于正啁啾时的特性，两种情况下的超连续谱特性均劣于无啁啾时的超连续谱特性，超连续谱特性劣化程度随啁啾参量$|C|$的增加而增大。

在凹形色散平坦光纤中，光脉冲频谱质心红移随啁啾参量 C 增加而增加，在方均根谱宽达到最大值之前，谱宽及其展宽速度随 C 值的增加而增加，在 DFF-CCDP 中的超连续谱特性优于凸形色散平坦光纤中的超连续谱特性。$|C|\leqslant 1$ 情况下，负啁啾时的最佳超连续谱特性优于正啁啾时的谱特性。入射光脉冲宽度越窄、色散参量 D_0 越大，产生的超连续谱越宽，获得最佳超连续谱所需光纤长度越短。超连续谱的宽度随参量 k_1 和 k_2 增大而减小，获得最佳超连续谱所需光纤长度随之而增加。入射光脉冲特性参量和光纤参量过小或过大，产生的超连续谱的整体特性较差。光纤中产生超连续谱和优化光纤参量设计中，应根据实际情况，综合考虑选择不同的参量。

参 考 文 献

[1] Agrawal G P. Nonlinear Fiber Optics. 5th edition. Singapore: Elsevier Pte Ltd., 2012.

[2] Sotobayashi H, Chujo W, Konishi A, et al. Wavelength-band generation and transmission of 3.24-Tbit/s (81-channel×40-Gbit/s) carrier-suppressed return-to-zero format by use of a single supercontinuum source for frequency standardization. J. Opt. Soc. Am. B, 2002, 19(11): 2803～2809.

[3] Ohara T, Takara H, Yamamoto T, et al. Over-1000-channel ultradense wdm transmission with supercontinuum multicarrier source. Journal of Lightwave Technology, 2006, 24(6):2311～2317.

[4] Mori K, Takara H, Kawanishi S, et al. Flatly broadened supercontinuum generation in a dispersion decreasing fiber with convex dispersion profile. Electronics Letters, 1997, 33(21): 1806～1807.

[5] Nan Y B, Lou C Y, Wang J P, et al. Improving the performance of a multiwavelength continuous-wave optical source based on supercontinuum by suppressing degenerate four-wave mixing. Optics Communications, 2005, 256: 428～434.

[6] Yang J W, Chae C J. WDM-PON upstream transmission using Fabry-Perot laser diodes externally injected by polarization-insensitive spectrum-sliced supercontinuum pulses. Optics Communications, 2006, 260: 691～695.

[7] Smirnov S V, Ania-Castanon J D, Ellingham T J, et al. Optical spectral broadening and supercontinuum generation in telecom applications. Optical Fiber Technology, 2006, 12:122～147.

[8] 娄采云, 李玉华, 伍剑, 等. 利用 10GHz 主动锁模光纤激光器在 DSF 中产生超连续谱. 中国激光, 2000, 27(9): 814～818.

[9] 伍剑, 李玉华, 娄采云, 等. 利用超连续谱光源产生超短光脉冲.光学学报, 2000, 20(3): 325～329.

[10] 王肇颖, 贾东方, 葛春风, 等. 10 GHz 再生锁模光纤激光器获得光纤超连续谱的研究. 光电子·激光, 2006, 17(3): 9～13.

[11] Szkulmowski M, Wojtkowski M, Bajraszewski T, et al. Quality improvement for high resolution in vivo images by spectral domain optical coherence tomography with supercontinuum source. Optics Communications, 2005, 246: 569～578.

[12] Kano H, Hamaguchi H. Femtosecond coherent anti-Stokes Raman scattering spectroscopy using supercontinuum generated from a photonic crystal fiber. Applied Physics Letters, 2004, 85(19): 4298～4300.

[13] Lindfors K, Kalkbrenner T, Stoller P, et al. Detection and spectroscopy of gold nanoparticles using supercontinuum white light confocal microscopy. Physical Review Letters, 2004, 93(3): 037401.

[14] Wang H C, Alismail A, Barbiero G, et al. Cross-polarized, multi-octave supercontinuum generation. Opt. Lett., 2011, 42: 2595～2598.

[15] Petersen C R, Engelsholm R D, Markos C, et al. Increased mid-infrared supercontinuum bandwidth and average power by tapering large-mode-area chalcogenide photonic crystal fibers. Opt. Express, 2017, 25: 15336～15348.

[16] Yin K, Zhang B, Yang L Y, et al. 15.2 W spectrally flat all-fiber supercontinuum laser source

with >1W power beyond 3.8μm. Opt. Lett., 2011, 42: 2334～2337.

[17] Valliammai M, Sivabalan S. Wide-band supercontinuum generation in mid-IR using polarization maintaining chalcogenide photonic quasi-crystal fiber. Appl. Opt., 2017, 56: 4797～4806.

[18] Khalifa A B, Salem A B, Cherif R. Mid-infrared supercontinuum generation in multimode As_2Se_3 chalcogenide photonic crystal fiber. Appl. Opt., 2017, 56: 4319～4324.

[19] Eftekhar M A, Wright L G, Mills M S, et al. Versatile supercontinuum generation in parabolic multimode optical fibers. Opt. Express, 2017, 25: 9078～9087.

[20] Zhao S L, Yang H, Zhao C J, et al. Harnessing rogue wave for supercontinuum generation in cascaded photonic crystal fiber. Opt. Express, 2017, 25: 7192～7202.

[21] Kedenburg S, Strutynski C, Kibler B, et al. High repetition rate mid-infrared supercontinuum generation from 1.3 to 5.3 μm in robust step-index tellurite fibers. J. Opt. Soc. Am. B, 2017, 34: 601～607.

[22] Strutynski C, Froidevaux P, Désévédavy F, et al. Tailoring supercontinuum generation beyond 2 μm in step-index tellurite fibers. Opt. Lett., 2017, 42: 247～250.

[23] Tarnowski K, Martynkien T, Mergo P, et al. Coherent supercontinuum generation up to 2.2 μm in an all-normal dispersion microstructured silica fiber. Opt. Express, 2016, 24: 30523～30536.

[24] Mori K, Takara H, Kawanishi S. Analysis and design of supercontinuum pulse generation in a single-mode optical fiber. J. Opt. Soc. Am. B, 2001, 18(12): 1780～1792.

[25] 娄采云, 高以智, 王建萍, 等. 光纤中超连续谱产生的理论与实验研究. 清华大学学报(自然科学版), 2003, 43(4): 441～445.

[26] 贾东方, 丁永奎, 胡志勇, 等. 光纤中超连续谱产生机理研究. 光电子·激光, 2004, 15(5): 612～620.

[27] 王肇颖, 王永强, 李智勇, 等. 皮秒脉冲在色散位移光纤中产生的超连续谱. 光电子·激光, 2004, 15(5): 528～533.

[28] 李智勇, 王肇颖, 王永强, 等. 基于 100m 色散位移光纤的超连续谱实验研究. 光子学报, 2004, 33(9): 1064～1066.

[29] 成纯富, 王晓方, 鲁波. 飞秒光脉冲在光子晶体光纤中的非线性传输和超连续谱产生. 物理学报, 2004, 53(6): 1826～1830.

[30] 陈泳竹, 李玉忠, 屈圭, 等. 反常色散平坦光纤产生平坦宽带超连续谱的数值研究. 物理学报, 2006, 55(2): 717～722.

[31] 郑宏军, 刘山亮, 黎昕, 等. 初始啁啾对凸形色散平坦光纤超连续谱的影响. 光电子·激光, 2007, 18(8): 919～923.

[32] Li X, Zheng H J, Xia Y J. Supercontinuum spectrum generation in dispersion-flatted fiber with concave dispersion profile. Proc. of SPIE, 2008, 7136: 71363A-1-6.

第 7 章　啁啾脉冲拉曼放大及其增益系数测量

拉曼放大是基于受激拉曼散射原理，以光纤作为增益介质而实现的全光放大。相对于稀土掺杂的光纤放大而言,它具有更大的增益带宽、灵活的增益谱区、更低的放大器自发辐射噪声以及能够有效抑制信噪比(SNR)的劣化等优点，近年来在光纤传输系统中获得越来越多的应用[1-17]。拉曼放大光孤子传输系统的研究中大多采用自相关技术等测量脉冲时域变化，难于准确判断脉冲的时域波形[2-7]。近几年发展起来的二次谐波频率分辨光学门(SHG-FROG)测量技术[18-39]可以有效抑制背景，具有较高的动态范围，能够测量光脉冲的脉宽、谱宽、波形、相位等特征参数信息，可以广泛应用于各种激光脉冲的测量。鉴于以往实验条件等各种因素的限制，本章采用能准确测量脉冲时域波形的 SHG-FROG 脉冲分析仪研究了拉曼放大对光孤子传输特性的影响，并与理论预期作了对比，为拉曼放大器及其通信系统的优化设计提供了重要依据；考虑到光纤拉曼放大在高速宽带光通信中有重要作用、拉曼放大增益系数是设计和实现光纤拉曼放大的一个重要指标，本章同时提出通过采用 SHG-FROG 新技术测量啁啾孤子脉冲的时域波形、脉冲宽度、相位等参量，给出了不同于已有文献[1，40，41]测量拉曼增益系数的一种新方法。

7.1　啁啾脉冲在拉曼放大系统中的传输特性

7.1.1　拉曼放大的实验装置

拉曼放大的实验装置如图 7-1 所示。

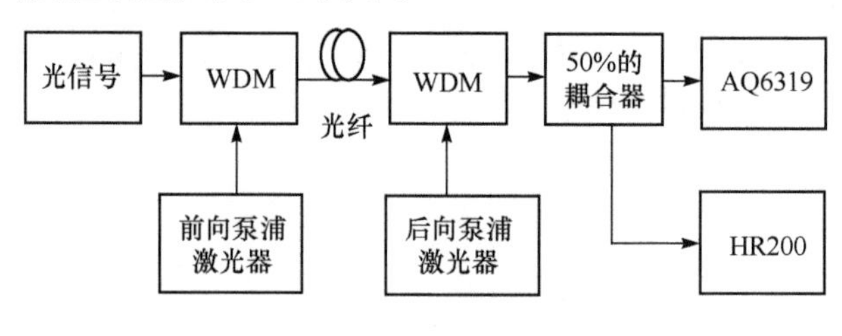

图 7-1　实验装置

图 7-1 中光信号可以是连续光信号，或者是小信号脉冲，或者是孤子脉冲，WDM 是波分复用器，将泵浦激光器产生的泵浦波和光信号耦合到光纤，在光纤中实现泵浦波对光信号的拉曼放大，50%的耦合器(50% Coupler)将光信号平均分成两路，分别进入光谱仪 AQ6319 和脉冲分析仪 HR200。实验中所用光纤是 G.652 标准单模光纤。

7.1.2　拉曼放大的实验结果与讨论

7.1.2.1　拉曼泵浦源特性及其自发拉曼谱

实验中采用的两支拉曼泵浦半导体激光器(连续泵浦波)中心波长分别为 1440nm 和 1450nm，它们的泵浦光功率与供电电流的关系如图 7-2 所示。横坐标是供电电流，单位是安培(A)，纵坐标是光功率，单位是分贝毫瓦(dBm)。当电流 $I=0.1$A 时，两个激光器均处于开启状态，输出大约 14dBm 的光功率。此后，两个激光器的输出光功率随电流的增加而缓慢增加。对应相同的电流，1450nm 激光器输出的光功率比 1440nm 激光器输出的功率稍大。

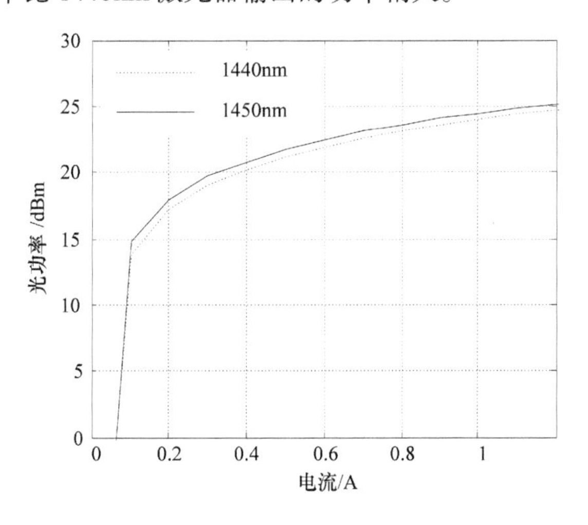

图 7-2　泵浦光功率与供电电流的关系

图 7-3 所示是在泵浦电流为 1A 时 1440nm 和 1450nm 泵浦源分别经过 25km 标准单模光纤形成的自发拉曼谱。实线所示对应 1440nm 泵浦源前向泵浦的情况，虚线所示对应 1440nm 泵浦源后向泵浦的情况，点划线所示对应 1450nm 泵浦源前向泵浦的情况，点线所示对应 1450nm 泵浦源后向泵浦的情况。横坐标是波长，单位是纳米(nm)，纵坐标是放大器自发辐射(ASE)功率，单位是分贝毫瓦(dBm)。

由图可见，当泵浦功率相同时，1440nm 后向泵浦自发拉曼谱的功率比前向泵浦时大 2dB 左右；两者–3dB 带宽都从 1521nm 至 1554nm，约 33nm。当泵浦功率增加时，1440nm 的后向泵浦和前向泵浦自发拉曼谱的功率随之增加；–3dB 带宽随之变化不大，仍然保持大约 33nm。

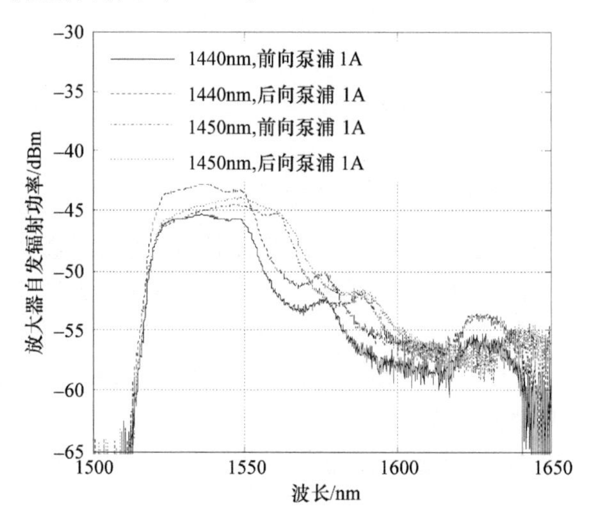

图 7-3　在泵浦电流为 1A 时，1440nm 和 1450nm 泵浦源分别经过 25km 标准单模光纤形成的自发拉曼谱

当泵浦功率相同时，1450nm 后向泵浦自发拉曼谱的功率比前向泵浦时稍大；两者–3dB 带宽都从 1521nm 至 1565nm，大约 44nm，比 1440nm 泵浦时的 33nm 宽得多，且向长波长方向延伸。当泵浦功率增加时，1450 nm 的后向泵浦和前向泵浦自发拉曼谱的功率随之增加；–3dB 带宽随之变化不大，仍然保持 44nm 左右。可见，后向泵浦优于前向泵浦，1450nm 泵浦源优于 1440nm 的情况。

7.1.2.2　拉曼放大对孤子传输特性的影响

1) 输入孤子脉冲的特性

半导体锁模激光器 TMLL1550 输出的 10GHz、1550nm 脉冲经 KPS 掺铒光纤放大器放大并获得光孤子脉冲。在孤子脉冲传输前，采用 SHG-FROG 脉冲分析仪对孤子脉冲进行了实验测量分析，将孤子脉冲波形和相位数据由 SHG-FROG 脉冲分析仪导入作者设计的 Matlab 计算程序中进行曲线拟合，得到了实验中孤子脉冲的包络电场表达式和啁啾等参量。

图 7-4 所示是 468mW 输入孤子脉冲的时域波形(a)和相位曲线(b)。横坐标是时间，单位是皮秒(ps)，图(a)中纵坐标是孤子脉冲归一化强度，图(b)中纵坐标是

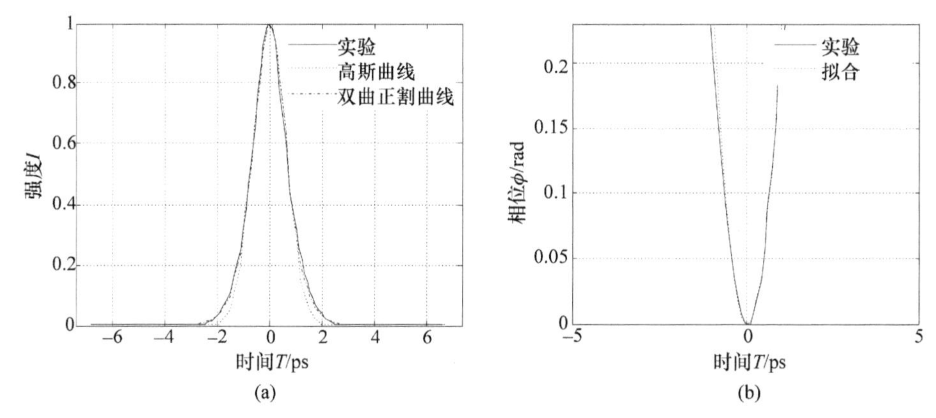

图 7-4　468mW 输入孤子脉冲的时域波形(a)和相位曲线(b)

实线所示是实验曲线，图(a)中点线和点划线分别是高斯和双曲正割曲线，图(b)中点线是方程(7-1)的相位曲线

脉冲相位，单位是弧度(rad)，实线所示是实验曲线，与双曲正割脉冲

$$u(0,\tau)=A\operatorname{sech}(\tau)\exp(-0.5\mathrm{i}C\tau^2) \tag{7-1}$$

的时域波形和相位曲线非常吻合。式(7-1)中，u 是包络电场，$A=1$ 是归一化振幅，$\tau=\dfrac{T}{T_0}$ 是归一化时间，$T_0=1.47/1.763\text{ps}$ 是脉冲半宽度，$C=-0.35$ 是线性啁啾参量。

2) 输出孤子脉冲的特性

考虑到上述自发拉曼谱情况，本节研究了连续泵浦波 1450nm 后向泵浦的拉曼放大对孤子传输特性的影响。在泵浦光是连续波的情况下，拉曼放大的有效光纤长度为

$$L_{\text{eff}}=\frac{1}{\alpha_{\text{p}}}[1-\exp(-\alpha_{\text{p}}L)] \tag{7-2}$$

式中，α_{p} 是泵浦光频率处的光纤损耗，在 1450nm 处，实验测得 $\alpha_{\text{p}}=0.29\text{dB/km}$。当光纤较长，$\alpha_{\text{p}}L\gg 1$ 时，$L_{\text{eff}}\approx\dfrac{1}{\alpha_{\text{p}}}=\dfrac{1}{0.29\times 4.343^{-1}}\approx 15\text{km}$。因此，作者选择了两条不同长度的光纤进行实验，一条光纤为 25.284km，其长度大于拉曼放大的有效光纤长度，模场直径为 9.24μm，1550nm 处的色散参量 $D=14.97\text{ps/(nm}\cdot\text{km)}$，光纤损耗为 0.182dB/km，色散斜率为 $k=0.086\text{ps/(nm}^2\cdot\text{km)}$；另一条光纤为 9km，其长度小于拉曼放大的有效光纤长度，模场直径为 9.07μm，1550nm 处的色散参量 $D=15.07\text{ps/(nm}\cdot\text{km)}$，光纤损耗为 0.188dB/km，色散斜率为 $k=0.086\text{ps/(nm}^2\cdot\text{km)}$。

将不同情况下的输出脉冲波形和相位数据由 SHG-FROG 脉冲分析仪导入作者设计的 Matlab 计算程序中进行曲线拟合，得到了实验中输出脉冲的时域波形等。图 7-5 所示是 468mW 时孤子脉冲在 9km 光纤中传输后的时域波形。图(a)为无拉曼放大，图(b)为拉曼泵浦电流为 0.5A，图(c)为拉曼泵浦电流为 1A，实线所示是实验波形，点线所示是高斯曲线，点划线所示是双曲正割曲线。该图与图 7-4(a)坐标、图例相同。

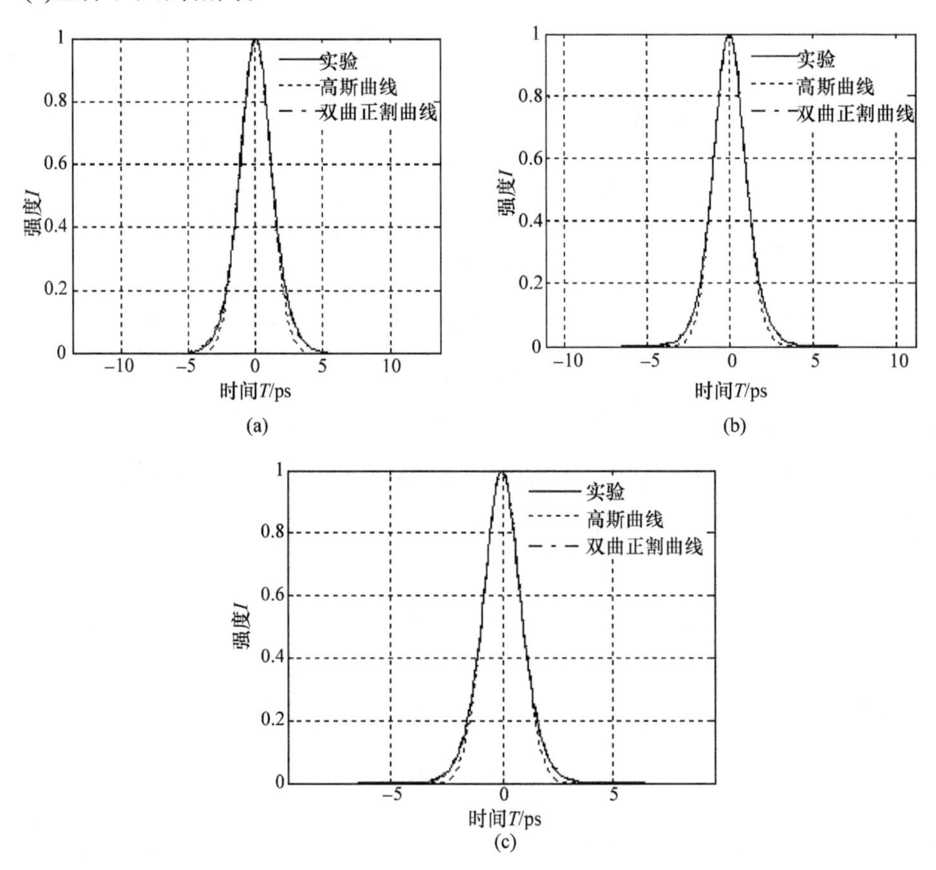

图 7-5　468mW 孤子脉冲在 9km 光纤中传输后的时域波形

(a)无拉曼放大；(b)拉曼泵浦电流为 0.5A；(c)拉曼泵浦电流为 1A

由图 7-4 和图 7-5 的实验结果比较可得，10GHz 孤子短脉冲经 9km 光纤传输后，孤子脉冲时域波形仍与双曲正割脉冲波形保持一致，拉曼放大能够压缩孤子脉冲、补偿光纤损耗。没有拉曼放大时，在光纤损耗的影响下，468mW 孤子脉冲经 9km 光纤传输后时域半峰全宽由输入脉宽 1.47ps 逐渐展宽为 2.712ps。拉曼泵浦电流为 1A 时，468mW 孤子脉冲输出时域半峰全宽是 1.934ps，比无拉曼放大时的输出脉冲宽度 2.712ps 明显减小，比拉曼泵浦电流为 0.5A 时的输出脉冲宽度 2.227ps 小，比输入脉宽 1.47ps 稍宽。可见，拉曼放大压缩了孤子脉冲，补偿了光纤损耗。

图 7-6 所示为 468mW 时孤子脉冲在 25.284km 光纤中传输后的时域波形。图(a)为无拉曼放大，图(b)为拉曼泵浦电流为 1A，实线所示是实验波形，点线所示是高斯曲线，点划线所示是双曲正割曲线。该图与图 7-4(a)中坐标、图例相同。

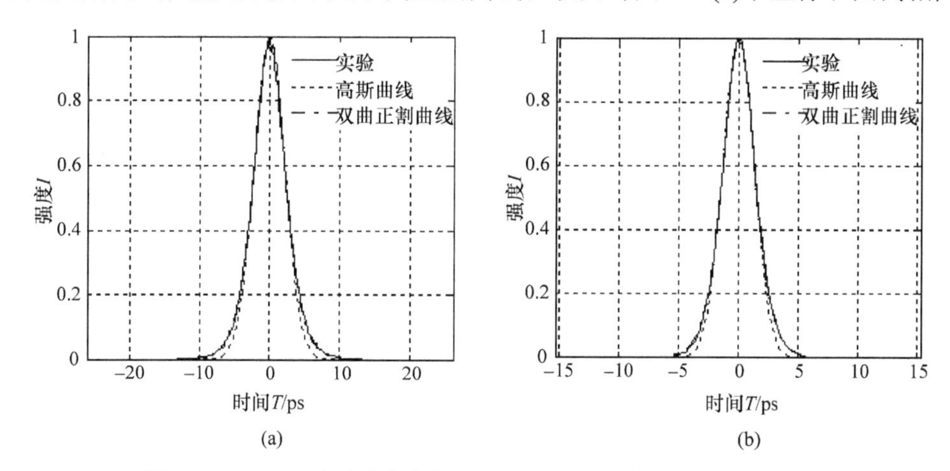

图 7-6　468mW 时孤子脉冲在 25.284km 光纤中传输后的时域波形
(a)无拉曼放大，(b)拉曼泵浦电流为 1A

由图 7-4～图 7-6 实验结果比较可得，在传输光纤长度为 25.284km 的情况下，没有拉曼放大时，10GHz 孤子短脉冲在 25.284km 光纤中传输后，孤子脉冲时域波形虽然仍与双曲正割脉冲波形保持一致，但是，在光纤损耗的影响下，孤子脉冲半峰全宽由输入脉宽 1.47ps 逐渐展宽为 5.191ps。在拉曼放大作用下，孤子短脉冲时域波形不仅仍与双曲正割脉冲波形保持一致，而且拉曼增益部分补偿了光纤损耗，输出孤子脉冲半峰全宽是 3.075ps，比无拉曼放大时的输出脉冲宽度 5.191ps 明显减小，是输入脉宽 1.47ps 的 2 倍多。

上述实验研究表明，拉曼放大能够压缩孤子脉冲、补偿光纤损耗，但不改变孤子脉冲的时域波形。在传输光纤长度小于拉曼放大有效光纤长度情况下，拉曼增益能够完全补偿光纤损耗；在传输光纤长度大于拉曼放大有效光纤长度情况下，拉曼增益能够部分补偿光纤损耗。拉曼放大对孤子脉冲的压缩和对光纤损耗的补偿能力与泵浦激光器特性有关，随实验中泵浦功率的增加而增大。

7.1.2.3　拉曼泵浦光偏振特性对孤子传输特性的影响

图 7-7 所示是 1450nm 后向泵浦光偏振特性对孤子传输特性的影响

(25.284km)。实线所示是调节泵浦光偏振得到的孤子脉冲最大平均功率，点线所示是调节泵浦光偏振得到的孤子脉冲最小平均功率。横坐标是拉曼泵浦电流，单位是安培(A)，纵坐标是信号功率，单位是分贝毫瓦(dBm)。由图可得，调节泵浦光偏振，光孤子脉冲最大平均功率和最小平均功率之差大约为 0.2dBm。相同实验条件下，而将孤子脉冲更换为小信号脉冲时，调节泵浦光偏振，小信号脉冲最大平均功率和最小平均功率之差大约为 1dBm。将孤子脉冲更换为小信号连续光时，调节泵浦光偏振，小信号连续光最大平均功率和最小平均功率之差大约为 2dBm。实验研究表明，光孤子脉冲对拉曼放大泵浦光偏振特性不敏感，这与文献[42]和[43]中光孤子在光纤中保持均一偏振态的理论一致。

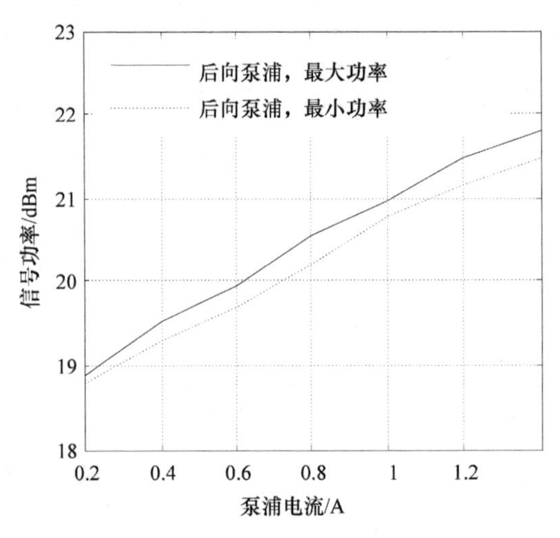

图 7-7　泵浦光偏振特性对孤子传输特性的影响

7.1.3　拉曼放大的数值计算与分析

光纤拉曼放大中，光纤损耗的拉曼增益系数定义为

$$\alpha_{\mathrm{R}} = g\exp[-\alpha_{\mathrm{p}}(L-z)] \tag{7-3}$$

式中，α_{p} 是在泵浦波长的光纤损耗系数，L 为光纤长度，z 为光纤传输变量，参量 g 与泵浦波的光功率有关，考虑到在泵浦处(z=0)的拉曼增益能够完全抵消光纤损耗，本节中令 $g=\alpha_{\mathrm{s}}$，α_{s} 是在信号波长的光纤损耗系数。考虑分布式拉曼放大对光纤损耗的补偿效应，忽略广义非线性薛定谔方程(1-21)中的高阶色散项和高阶非线性，光孤子在光纤中传输的归一化方程修正为

$$\mathrm{i}\frac{\partial u}{\partial \xi}+\frac{1}{2}\frac{\partial^2 u}{\partial \tau^2}+|u|^2 u=-0.5\mathrm{i}(\alpha_{\mathrm{s}}-\alpha'_{\mathrm{R}})L_D u \tag{7-4}$$

式中，ξ 是传输距离，归一化到色散长度；$\tau = T/T_0$ 为归一化时间；$\alpha_R' = \alpha_s \exp[-\alpha_p(L'-\xi)L_D]$ 为 α_R 归一到色散长度 $L_D = T_0^2 / |\beta_2|$ 的参量。下面以式(7-1)为输入脉冲，采用分步傅里叶方法数值研究拉曼放大对光孤子传输特性的影响。传输距离9km和25.284km分别归一化为248.6和693.9个色散长度。

图7-8所示是数值计算得到的468mW孤子时域宽度随传输距离(9km)的变化。虚线所示是考虑光纤损耗的情况，实线所示是考虑光纤损耗和拉曼增益的情况，实点和圆圈是对应上述情况的实验数据。横坐标是传输距离，归一化到色散长度；纵坐标是孤子脉冲时域半峰全宽，单位是皮秒(ps)。

由上图可得，在仅考虑光纤损耗情况下，具有初始负啁啾的输入孤子脉冲随传输距离增加周期性展宽，在248.6个色散长度(9km)处的脉冲半峰全宽2.675ps与实验数据2.712ps基本吻合。考虑光纤损耗和拉曼增益情况下，输入孤子脉冲随传输距离增加周期性展宽速度减小，在248.6个色散长度处的脉冲半峰全宽2.01ps与实验数据1.934ps基本一致，与无光纤损耗和无拉曼增益情况下的数值结果1.804ps基本吻合。这表明，在传输光纤长度小于拉曼放大有效光纤长度情况下拉曼增益完全补偿了光纤损耗。

图7-9所示是数值计算得到的468mW孤子时域宽度随传输距离(25.284km)的变化。本图与图7-8中坐标、图例相同。由图7-9可得，仅考虑光纤损耗情况下，输入孤子脉冲半峰全宽随传输距离增加逐渐展宽到693.9个色散长度(25.284km)处的5.36ps，与实验测量得到的5.191ps基本吻合。考虑光纤损耗和拉曼增益情况

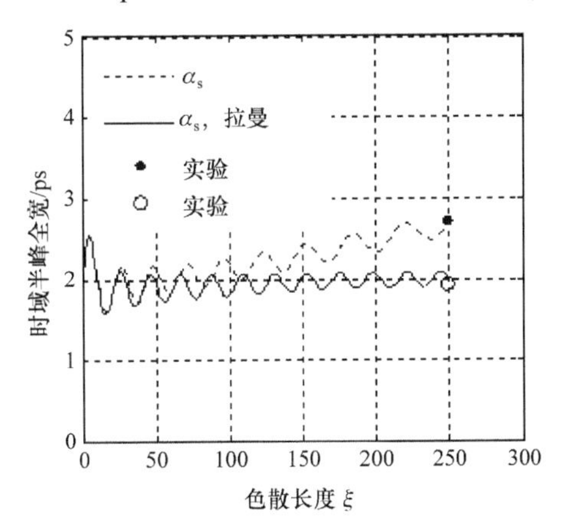

图7-8　468mW孤子时域宽度随传输距离(9km)的变化

虚线所示是考虑光纤损耗的情况，实线所示是考虑光纤损耗和拉曼增益的情况，实点和圆圈是对应两种情况的实验数据

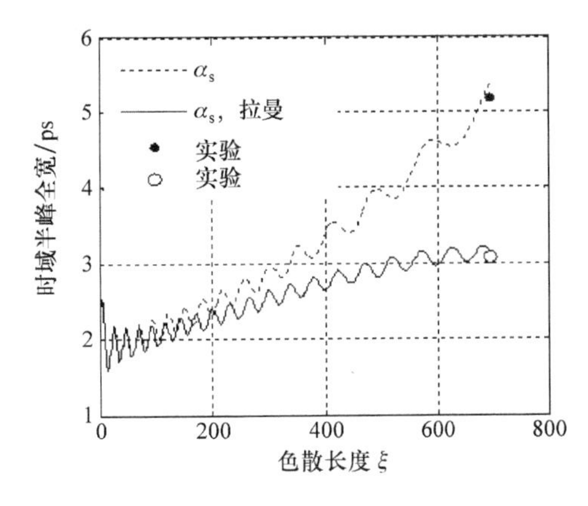

图7-9　468mW孤子时域宽度随传输距离(25.284km)的变化

虚线所示是考虑光纤损耗的情况，实线所示是考虑光纤损耗和拉曼增益的情况，实点和圆圈是对应两种情况的实验数据

下，输入孤子脉冲随传输距离增加周期性展宽速度明显减小，在 693.9 个色散长度处的脉冲半峰全宽 3.158ps 与实验数据 3.075ps 基本一致，是无光纤损耗和无拉曼增益情况下数值结果 1.811ps 的 1.7 倍，是输入脉宽 1.47ps 的 2 倍多。这表明，在传输光纤长度大于拉曼放大有效光纤长度情况下，拉曼增益能够部分补偿光纤损耗。

图 7-10 和图 7-11 所示分别是 468mW 时孤子脉冲在 9km 和 25.284km 光纤中传输后的时域波形。图(a)为无拉曼放大，图(b)为拉曼泵浦电流为 1A，实线所示是实验波形，虚线所示是数值计算得到的时域波形，点线所示是高斯曲线，点划线所示是双曲正割曲线。两图与图 7-4(a)中坐标相同。

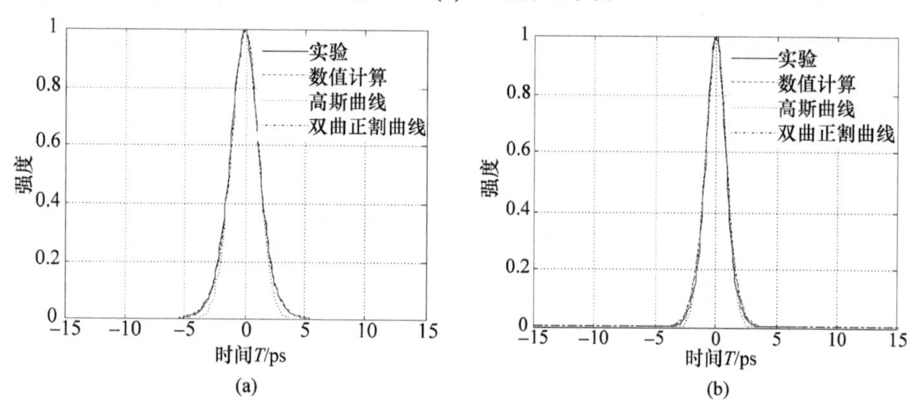

图 7-10　468mW 孤子脉冲在 9km 光纤中传输后的时域波形
(a)无拉曼放大，(b)拉曼泵浦电流为 1A；实线所示是实验波形，虚线所示是数值计算得到的时域波形，点线所示是高斯曲线，点划线所示是双曲正割曲线

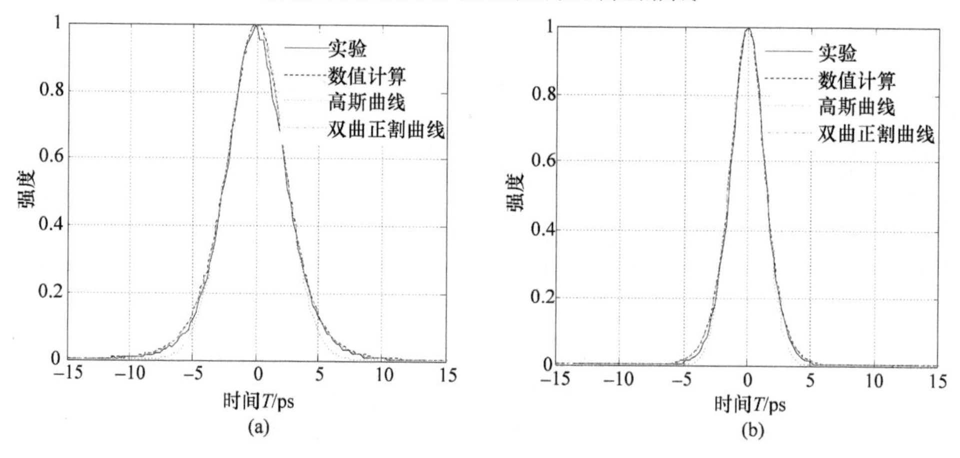

图 7-11　468mW 孤子脉冲在 25.284km 光纤中传输后的时域波形
(a)无拉曼放大，(b)拉曼泵浦电流为 1A，实线所示是实验波形，虚线所示是数值计算得到的时域波形，点线所示是高斯曲线，点划线所示是双曲正割曲线

由图可得，数值计算得到的孤子脉冲传输后时域波形与实验波形基本一致，均与双曲正割曲线吻合。这表明孤子脉冲在光纤损耗和拉曼增益作用下仍然保持双曲正割波形，并根据能量损耗或补偿的多少具有不同的脉冲宽度。

7.1.4 讨论与结论

本节修正了拉曼放大作用下的孤子脉冲传输方程，研究了拉曼放大对光孤子传输特性的影响，实验数据与采用分步傅里叶方法的数值研究结果一致。拉曼放大能够压缩孤子脉冲、补偿光纤损耗，但不改变孤子脉冲的时域波形。在传输光纤长度小于拉曼放大有效光纤长度的情况下，拉曼放大能够完全补偿光纤损耗；在传输光纤长度大于拉曼放大有效光纤长度的情况下，拉曼放大能够部分补偿光纤损耗。拉曼放大对孤子脉冲的压缩和对光纤损耗的补偿能力与泵浦激光器特性有关，随实验中泵浦功率的增加而增大。光孤子脉冲对拉曼放大泵浦光偏振特性不敏感，与光孤子在光纤中保持均一偏振态的理论一致。

7.2 啁啾脉冲在拉曼放大系统中的增益系数测量

考虑到光纤拉曼放大在国际科技界和高速宽带光通信中有重要作用，并且拉曼放大增益系数是设计和实现光纤拉曼放大的一个重要指标，本节提出通过采用 SHG-FROG 新技术测量啁啾孤子脉冲的时域波形、脉冲宽度、相位等参量，给出了不同于已有文献[1，40，41]测量拉曼增益系数的一种新方法。

7.2.1 拉曼放大的系数测量原理

对于连续泵浦波的情况，考虑连续泵浦波和信号光波(斯托克斯波)之间的相互作用应遵循下列耦合方程[1]

$$\frac{\mathrm{d}I_{\mathrm{s}}}{\mathrm{d}z}=g_{\mathrm{R}}I_{\mathrm{p}}I_{\mathrm{s}}-\alpha_{\mathrm{s}}I_{\mathrm{s}} \tag{7-5}$$

$$\frac{\mathrm{d}I_{\mathrm{p}}}{\mathrm{d}z}=-\frac{\omega_{\mathrm{p}}}{\omega_{\mathrm{s}}}g_{\mathrm{R}}I_{\mathrm{p}}I_{\mathrm{s}}-\alpha_{\mathrm{p}}I_{\mathrm{p}} \tag{7-6}$$

其中，I_{s}为信号光波光强，I_{p}为泵浦波光强，g_{R}为拉曼增益系数，α_{s}和 α_{p}分别为信号光波频率和泵浦频率处的光纤损耗，ω_{p}和ω_{s}分别为泵浦光波和信号光波的频率。

式(7-6)中等号右边的第一项代表泵浦损耗，由于信号光强总小于泵浦光强，故泵浦损耗可忽略。在忽略掉式(7-6)右边第一项(泵浦消耗)的条件下，联立方程(7-5)和(7-6)可求出拉曼放大输出端 $z=L$ 处的信号光波强度：

$$I_S(L)=I_s(0)\exp(g_R I_0 L_{eff}-\alpha_s L) \tag{7-7}$$

式中，I_0是 z=0 处的入射泵浦光强，$I_s(0)$是 z=0 处的信号光波的入射光强，

$$L_{eff}=[1-\exp(-\alpha_p L)]/\alpha_p \tag{7-8}$$

是有拉曼放大时的有效光纤长度。无拉曼泵浦时，输出端 $z=L$ 处的信号光波强度

$$I_s(L)=I_S(0)\exp(-\alpha_s L) \tag{7-9}$$

考虑到拉曼放大主要源于信号光波频率附近的拉曼增益，由式(7-7)和式(7-9)可得拉曼放大增益为

$$G_R=\frac{P_S}{P_s}=\exp(g_R P_0 L_{eff}/A_{eff}) \tag{7-10}$$

P_S是有拉曼泵浦时输出端 $z=L$ 处的信号光波功率，P_s是无拉曼泵浦时输出端 $z=L$ 处的信号光波功率，P_0是输入端泵浦光波的功率，$A_{eff}=\pi a^2$是光纤的有效纤芯面积，a为光纤的模场半径。由式(7-10)可得拉曼放大增益系数

$$g_R=\frac{KA_{eff}\ln(P_S/P_s)}{P_0 L_{eff}} \tag{7-11}$$

式中，K 为作者考虑不同泵浦功率时啁啾孤子脉冲的时域宽度因子，这是其他拉曼增益系数测量方法中所没有考虑的因素。

7.2.2　拉曼增益系数的实验测量

拉曼增益系数的实验测量装置如图 7-12 所示，TMLL1550 为可调谐半导体锁模脉冲激光器，ISO 为光隔离器，KPS EDFA 为可调增益掺铒光纤放大器，HR200 为二次谐波频率分辨光学门(SHG-FROG)脉冲分析仪，AQ6319 是光谱分析仪，WDM 是波分复用器，Raman1450 是拉曼泵浦光源。实验中所用光纤是 G.652 标准单模光纤，光纤长度为 9km，模场直径为 9.07μm，1550nm 处的色散参量 D=15.07ps/(nm · km)，光纤损耗为 0.188dB/km，色散斜率为 k=0.086ps/(nm^2 · km)。

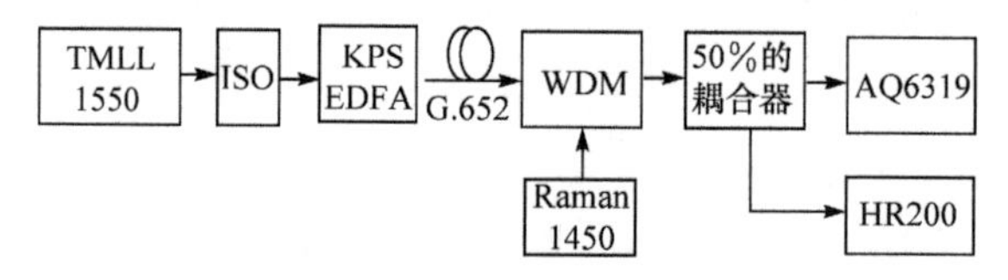

图 7-12　实验测量装置

TMLL1550 产生 10GHz 啁啾脉冲，经 ISO 由 KPS EDFA 放大得到的啁啾孤子脉冲信号进入 G.652 标准单模光纤；WDM 将 Raman1450 产生的 1450nm 泵浦光波耦合到 G.652 标准单模光纤，从而实现对啁啾孤子脉冲信号的后向拉曼放大；50%的耦合器将拉曼放大后的啁啾孤子脉冲信号平均分成两路，分别进入光谱仪 AQ6319 和脉冲分析仪 HR200 进行谱域和时域的实验测量。

7.2.2.1　拉曼泵浦源的特性

实验中采用 1450nm 的半导体激光器作为拉曼泵浦光源，拉曼泵浦光源特性如图 7-13 所示。图 7-13(a)是泵浦光功率随泵浦电流的变化，图 7-13(b)为泵浦电流为 1A 时 1450nm 泵浦光源经过标准单模光纤形成的自发拉曼谱。拉曼泵浦激光器的输出光功率随泵浦电流的增加逐渐增加。自发拉曼谱的–3dB 带宽约为 44nm。当泵浦功率增加时，自发拉曼谱的功率随之增加，–3dB 带宽随之变化不大。

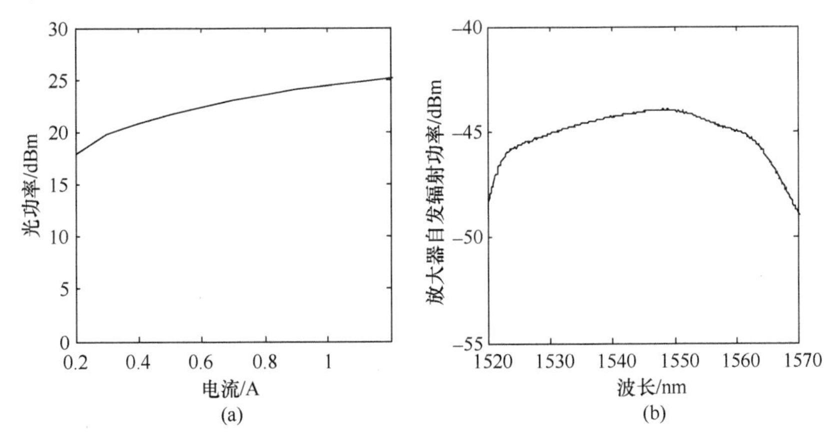

图 7-13　拉曼泵浦光源的特性

(a)泵浦光功率随泵浦电流的变化；(b)泵浦电流为 1A 时 1450nm 泵浦光源经过标准单模光纤形成的自发拉曼谱

7.2.2.2　输入孤子脉冲的特性

按照图 7-12 实验装置，首先采用 SHG-FROG 脉冲分析仪和 AQ6319 光谱仪对由 KPS EDFA 放大得到的啁啾孤子脉冲信号进行实验测量。将 SHG-FROG 脉冲分析仪实验测量数据导入我们研制的 Matlab 计算程序中进行曲线拟合，得到了实验输入孤子脉冲波形、脉宽和啁啾等参量。

图 7-14 所示是啁啾孤子脉冲输入 G.652 标准单模光纤前的时域波形(a)、相位曲线(b)和光谱(c)，实线所示是实验曲线，与双曲正割脉冲

$$u(0,\tau)=\operatorname{sech}\tau\exp(-0.5\mathrm{i}C\tau^2) \tag{7-12}$$

的时域波形非常吻合。式(7-12)中，u 是电场的归一化复振幅(相对于脉冲峰值功率)，$\tau=T/T_0$ 是归一化时间，T_0＝1.409/1.763ps 是脉冲半宽度，脉冲的半峰全宽 T_{FWHM}＝1.409ps，C=−0.28 是线性啁啾参量，$\mathrm{i}=\sqrt{-1}$。从光谱可以看出，此为 10GHz 锁模啁啾孤子脉冲，其中心波长在 1550nm 附近，在中心波长附近光谱的对称性较好，其光功率为 525mW。

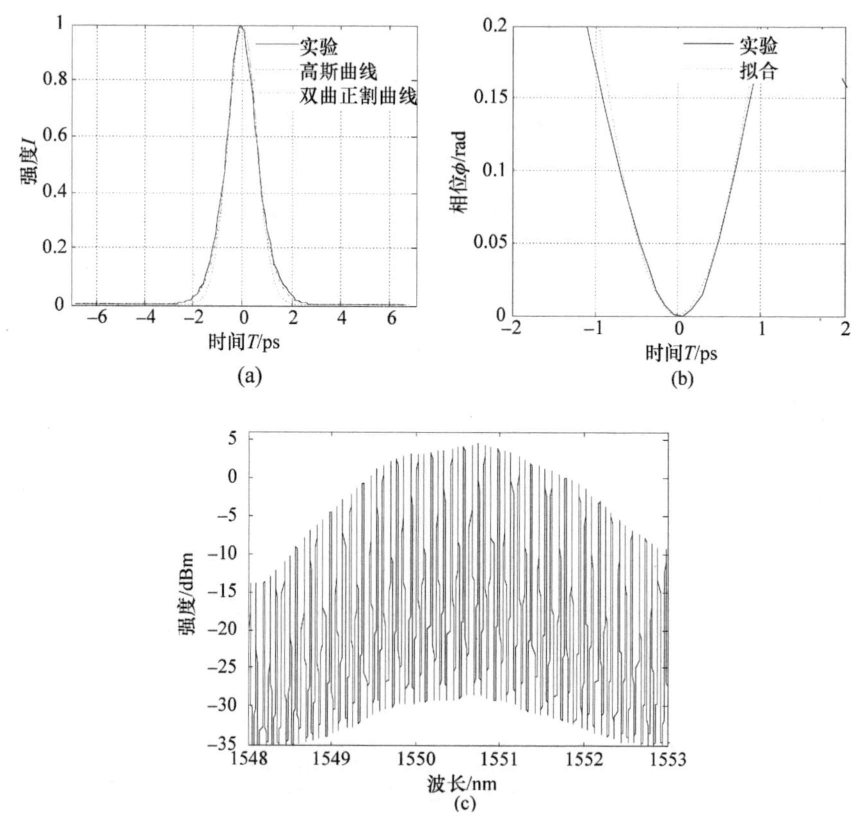

图 7-14　啁啾孤子脉冲的输入时域波形(a)、相位曲线(b)和光谱(c)

(b)点线是方程(7-12)的相位曲线

7.2.2.3　输出孤子脉冲的特性

通过研究连续泵浦波 1450nm 后向泵浦的拉曼放大对 1550 nm 啁啾孤子传输特性的影响，给出信号光波在 1550 nm 附近的拉曼增益系数。在泵浦光是连续波的情况下，光纤拉曼放大的有效光纤长度由式(7-8)得到

$$L_{\mathrm{eff}}=\frac{1}{\alpha_{\mathrm{p}}}[1-\exp(-\alpha_{\mathrm{p}}L)]\approx\frac{1}{\alpha_{\mathrm{p}}}=\frac{1}{0.29\times4.343^{-1}}\approx15(\mathrm{km})\tag{7-13}$$

式中，α_{p} 是泵浦光频率处的光纤损耗，在 1450nm 处，$\alpha_{\mathrm{p}}=0.29\mathrm{dB/km}$。当光纤较长，$\alpha_{\mathrm{p}}L\gg1$ 时，$L_{\mathrm{eff}}\approx1/\alpha_{\mathrm{p}}$。因此，这里选择了 9km 光纤进行实验。拉曼泵浦电流为 0.5A 时，拉曼泵浦光功率为 148.6mW；拉曼泵浦电流为 1A 时，拉曼泵浦光功率为 279.9mW。

图 7-15 所示是啁啾孤子脉冲在无拉曼放大时经 9km 光纤传输后的输出脉冲

的时域波形(a)和相位曲线(b)及光谱(c)，实线所示是实验曲线，与双曲正割脉冲式(7-12)的时域波形吻合，脉冲的时域半峰全宽 $T_{FWHM}=2.443$ps，$C=-0.35$ 是线性啁啾参量。由图 7-15 可以看出，啁啾孤子脉冲在无拉曼放大时经 9km 标准单模光纤传输后的输出脉冲的时域波形仍然保持双曲正割波形，但脉冲宽度比输入脉宽明显展宽，增加了大约 70%，线性啁啾稍有增加。从光谱可以得到，啁啾孤子脉冲经单模光纤传输后，在中心波长附近光谱仍然保持较好的对称性，比输入光谱稍有压缩，考虑 50%耦合器后其光功率为 208.9mW。

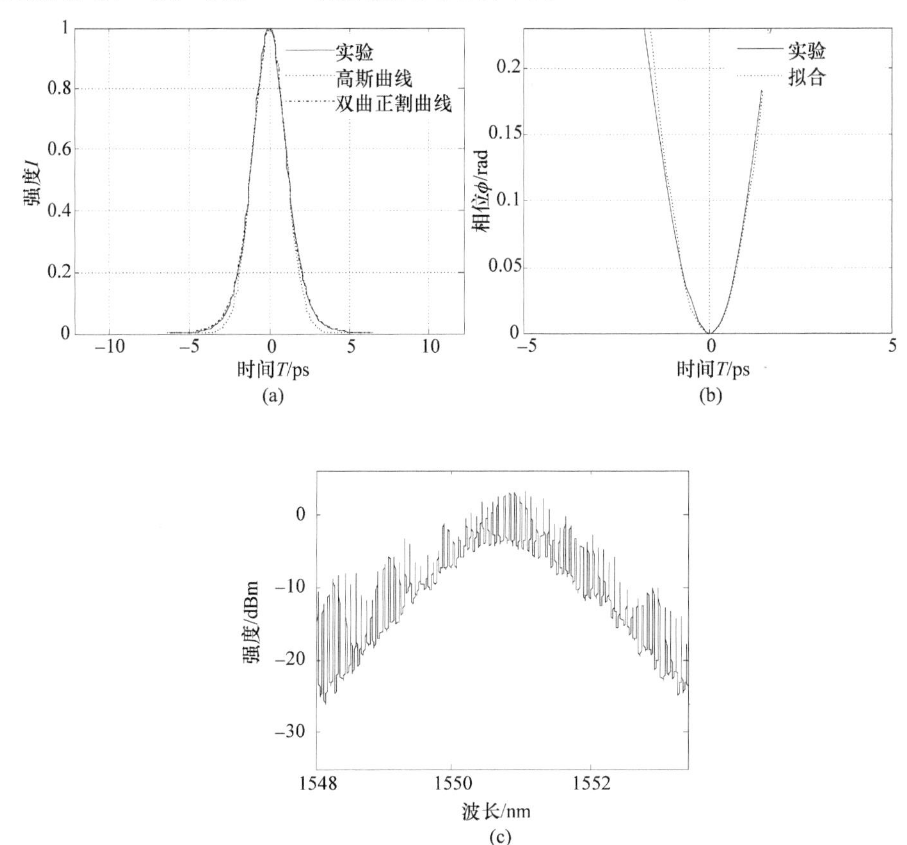

图 7-15　啁啾孤子脉冲在无拉曼放大时经 9km 光纤传输后的输出脉冲的时域波形(a)、相位曲线(b)和光谱(c)

(b)点线是方程(7-12)的相位曲线

图 7-16 所示是啁啾孤子脉冲在拉曼放大泵浦电流为 0.5A 时经 9km 光纤传输后的输出脉冲的时域波形(a)、相位曲线(b)和光谱(c)。实线所示是实验曲线，图(a)中点线和点划线分别是高斯和双曲正割曲线，图(b)中点线是方程(7-12)的相位曲线；脉冲的时域半峰全宽 $T_{FWHM}=2.125$ps，$C=-0.4$ 是线性啁啾参量。由

图 7-16 可以得到，啁啾孤子脉冲在有拉曼放大且泵浦电流为 0.5A 时经 9km 标准单模光纤传输后的输出脉冲的时域波形仍然保持双曲正割波形，但脉冲时域宽度比输入脉宽增加了大约 50%，比无拉曼放大时的输出孤子脉宽减小 20%，这表明拉曼增益压缩脉冲时域。此时线性啁啾比无拉曼放大时的输出孤子啁啾稍有增加。从光谱可以得到，有拉曼放大且泵浦电流为 0.5A 时，啁啾孤子脉冲在中心波长附近光谱仍然保持较好的对称性，比入射谱的对称性差；比无拉曼放大时的输出孤子光谱稍有展宽，考虑 50%耦合器后其光功率为 251.2mW。

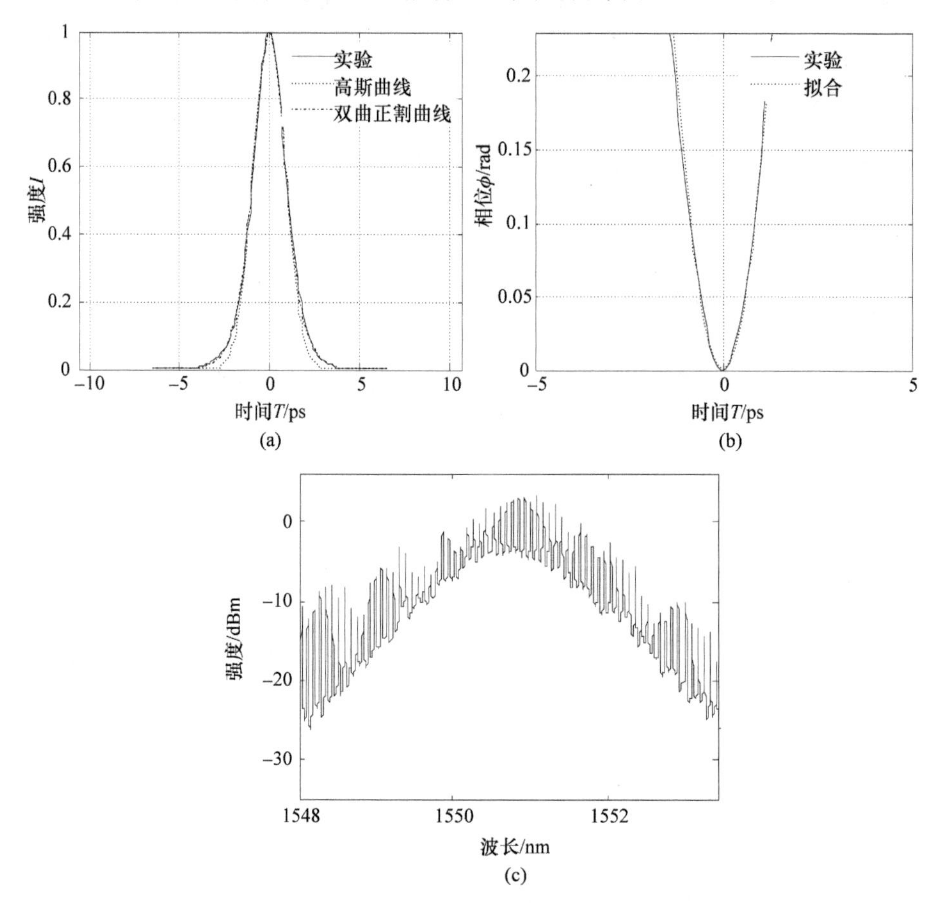

图 7-16　啁啾孤子脉冲在拉曼放大泵浦电流为 0.5A 时经 9km 光纤传输后的输出脉冲的时域波形(a)、相位曲线(b)和光谱(c)

图 7-17 所示是啁啾孤子脉冲在拉曼放大泵浦电流为 1A 时经 9km 光纤传输后的输出脉冲的时域波形(a)、相位曲线(b)和光谱(c)。实线所示是实验曲线，图(a)中点线和点划线分别是高斯和双曲正割曲线，图(b)中点线是方程(7-12)的相位

曲线；脉冲的时域半峰全宽 $T_{FWHM}=1.82ps$，$C=-0.4$ 是线性啁啾参量。由图 7-17 可以看出，啁啾孤子脉冲在有拉曼放大且泵浦电流为 1A 时经 9km 标准单模光纤传输后的输出脉冲的时域波形仍然保持双曲正割波形，脉冲时域宽度 1.82ps 比拉曼泵浦电流 0.5A 时的输出孤子脉冲宽度 2.125ps 小，比无拉曼放大时的输出孤子时域脉宽 2.443ps 明显减小，比输入脉宽 1.409ps 稍宽；这表明拉曼泵浦功率越大，拉曼增益越大，压缩脉冲时域越明显。此时线性啁啾比无拉曼放大时的输出孤子啁啾稍有增加。从光谱可以得到，有拉曼放大且泵浦电流为 1A 时，啁啾孤子脉冲在中心波长附近光谱仍然保持较好的对称性，比入射谱的对称性差；比有拉曼放大且泵浦电流 0.5A 时的输出孤子光谱稍有展宽，比无拉曼放大时的输出孤子光谱有明显展宽，考虑 50%耦合器后其光功率为 296mW。

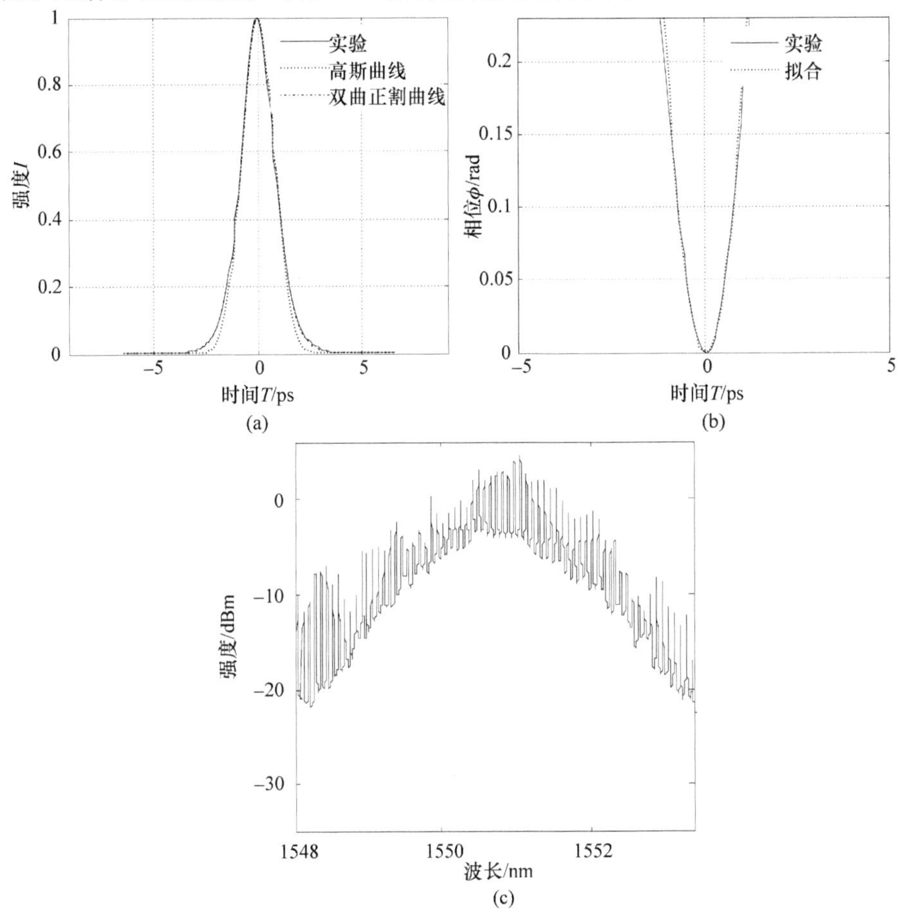

图 7-17　啁啾孤子脉冲在拉曼放大泵浦电流 1A 时经 9km 光纤传输后的输出脉冲的时域波形(a)、相位曲线(b)和光谱(c)

7.2.3　计算讨论与结论

考虑到不同泵浦功率时拉曼增益不同，啁啾孤子脉冲时域宽度不同，给出了啁啾孤子脉冲时域宽度因子 K，这是其他拉曼增益系数测量方法中所没有考虑的因素。有拉曼放大且泵浦电流 0.5A 时，$2\tau_0$=1.25，其中 2.125ps 是此时输出啁啾孤子脉冲的时域宽度，1.409ps 是输入啁啾孤子脉冲的时域宽度。有拉曼放大且泵浦电流为 1A 时，K=1.82/1.409，该系数说明拉曼放大中的光子能量只有一部分放大了孤子脉冲，没有完全转化到孤子脉冲中；这一点从输出啁啾孤子脉冲光谱对称性比输入光谱的对称性稍差、输入输出啁啾孤子脉冲时域宽度不同可以看出。传输光纤长度比拉曼放大有效长度小，故直接代入传输光纤长度 $L_{\text{eff}}=9\text{km}$；将有拉曼放大且泵浦电流为 0.5A 时的相关参数代入式(7-11)，得到拉曼增益系数

$$g_{\text{R1}}=\frac{KA_{\text{eff}}\ln(P_{\text{S}}/P_{\text{s}})}{P_0L_{\text{eff}}}=\frac{(2.125/1.409)\times3.14\times(9.07\times10^{-6}/2)^2\times\ln(251.2/208.9)}{148.6\times10^{-3}\times9\times10^3}$$
$$=1.38\times10^{-14}(\text{m/W})$$

将有拉曼放大且泵浦电流 1A 时的相关参数代入式(7-11)，得到拉曼增益系数

$$g_{\text{R2}}=\frac{KA_{\text{eff}}\ln(P_{\text{S}}/P_{\text{s}})}{P_0L_{\text{eff}}}=\frac{(1.82/1.409)\times3.14\times(9.07\times10^{-6}/2)^2\times\ln(296/208.9)}{279.9\times10^{-3}\times9\times10^3}$$
$$=1.18\times10^{-14}(\text{m/W})$$

则总拉曼增益系数为上述两个数据的平均：

$$g_{\text{R}}=(g_{\text{R1}}+g_{\text{R2}})/2=1.28\times10^{-14}(\text{m/W})$$

本节通过采用 SHG-FROG 技术测量啁啾孤子脉冲的时域波形、脉冲宽度、相位等参量，给出了不同于已有文献[1,40,41]测量拉曼增益系数的一种新方法。此方法得到的连续激光泵浦啁啾孤子脉冲的拉曼增益系数，与文献[40,41]中连续激光泵浦连续激光的拉曼增益系数在同一个数量级，比后者稍小。这是由于连续激光泵浦连续激光时，被拉曼放大的连续激光光谱较窄，一般远小于 1nm，较容易实现拉曼增益；而被拉曼放大的 10GHz 啁啾孤子脉冲的光谱较宽，–20dB 谱宽可达 5nm，每间隔 0.08nm(10GHz)一个锁模谱，各模式间需要满足一定的相位等条件，实现拉曼增益比连续激光泵浦连续激光情况较难。

7.3　本 章 小 结

本章修正了拉曼放大作用下的孤子脉冲传输方程，研究了拉曼放大对光孤子传输特性的影响，实验数据与采用分步傅里叶方法的数值研究结果一致；拉曼放大能够压缩孤子脉冲、补偿光纤损耗，但不改变孤子脉冲的时域波形；在传输光

纤长度小于拉曼放大有效光纤长度的情况下，拉曼放大能够完全补偿光纤损耗；在传输光纤长度大于拉曼放大有效光纤长度的情况下，拉曼放大能够部分补偿光纤损耗；拉曼放大对孤子脉冲的压缩和对光纤损耗的补偿能力与泵浦激光器特性有关，随实验中泵浦功率的增加而增大；光孤子脉冲对拉曼放大泵浦光偏振特性不敏感，与光孤子在光纤中保持均一偏振态的理论一致[44, 45]。

此外，本章还提出采用 SHG-FROG 技术测量啁啾孤子脉冲的时域波形、脉冲宽度、相位等参量，给出不同于已有文献[1, 40, 41]测量拉曼增益系数的一种新方法[46]。

参考文献

[1] Agrawal G P. Nonlinear Fiber Optics. 5th edition. Singapore: Elsevier Pte Ltd., 2012.

[2] Gouveia-Neto A S, Wigley P G J, Taylor J R. Soliton generation through Raman amplification of noise bursts. Optics Letters, 1989, 14(20): 1122～1124.

[3] Iwatsuki K, Suzuki K, Nishi S. Adiabatic soliton compression of gain-switched DFB-LD pulse by distributed fiber Raman amplification. IEEE Transactions Photonics Technology Letters, 1991, 3(12): 1074～1076.

[4] Murphy T E. 10-GHz 1.3-ps pulse generation using chirped soliton compression in a Raman gain medium. IEEE Photonics Technology Letters, 2002, 14(10): 1424～1426.

[5] Mollenauer L F, Stolen R H, Islam M N. Experimental demonstration of soliton propagation in long fibers: loss compensated by Raman gain. Optics Letters, 1985, 10(5): 229～231.

[6] Iwatsuki K, Nishi S, Saruwatari M, et al. 5Gb/s optical soliton transmission experiment using Raman amplification for fiber-loss compensation. IEEE Photonics Technology Letters, 1990, 2(7): 507～509.

[7] Hrimchuk A G, Onishchukov G, Lederer F. Long-haul soliton transmission at 1.3 μm using distributed Raman amplification. Journal of Lightwave Technology, 2001, 19(6): 837～841.

[8] Ereifej H N, Grigoryan V, Carter G M. 40 Gbit/s long-haul transmission in dispersion-managed soliton system using Raman amplification. Electronics Letters, 2001, 37(25): 1538～1539.

[9] Pincemin E, Hamoir D, Audouin O, et al. S. Distributed-Raman-amplification effect on pulse interactions and collisions in long-haul dispersion-managed soliton transmissions. J. Opt. Soc. Am. B, 2002, 19(5): 973～980.

[10] Tio A A B, Shum P. Propagation of optical soliton in a fiber Raman amplifier. Proceedings of SPIE, 2004, 5280: 676～681.

[11] Mollenauer L F, Smith K. Demonstration of soliton transmission over more than 4000 km in fiber with loss periodically compensated by Raman gain. Optics Letters, 1988, 13(8): 675～677.

[12] Chi S, Wen S. Interaction of optical solitons with a forward Raman pump wave. Optics Letters, 1989, 14(1): 84～86.

[13] Wen S, Wang T Y, Chi S. The optical soliton transmission amplified by bidirectional Raman pumps with nonconstant depletion. IEEE Journal of Quantum Electronics, 1991, 21(8): 2066～

2073.

[14] Levy G F. Raman amplification of solitons in a fiber optic ring. Journal of Lightwave Technology, 1996, 14(1): 72～76.

[15] 曹文华, 刘颂豪, 廖常俊, 等. 色散缓变光纤中的孤子效应拉曼脉冲产生. 中国激光, 1994, 21(6): 489～494.

[16] 李宏, 杨祥林, 刘堂坤. 暗孤子传输系统中调制拉曼泵浦的控制作用. 中国激光, 1997, 24(7): 654～658.

[17] 沈廷根, 郑浩, 李正华, 等. 掺杂光子晶体光纤的缺陷模增益谱与光孤子拉曼放大研究. 人工晶体学报, 2005, 34(6): 1065～1073.

[18] Kane D J, Trebino R. Characterization of arbitrary femtosecond pulses using frequency-resolved optical gating. IEEE J. Quantum Electron., 1993, 29(2): 571～579.

[19] DeLong K W, Trebino R, Hunter J, et al. Frequency-resolved optical gating with the use of second-harmonic generation. J. Opt. Soc. Am. B,1994, 11(11): 2206～2215.

[20] Gallmann L, Steinmeyer G, Sutter D H, et al. Collinear type II second-harmonic-generation frequency-resolved optical gating for the characterization of sub-10-fs optical pulses. Optics Letters, 2000, 25(4): 269～271.

[21] 王兆华, 魏志义, 滕浩, 等. 飞秒激光脉冲的谐波频率分辨光学开关法测量研究. 物理学报. 2003, 52(2): 362～366.

[22] 龙井华, 高继华, 巨养锋, 等. 用 SHG-FROG 方法测量超短光脉冲的振幅和相位. 光子学报. 2002, 31(10): 1292～1296.

[23] DeLong K W, Fittinghoff D N, Trebino R, et al. Pulse retrieval in frequency-resolved optical gating based on the method of generalized projections. Optics Letters, 1994, 19(24): 2152～2154.

[24] Hu J, Zhang G Z, Zhang B G, et al. Using frequency-resolved optical gating to retrieve amplitude and phase of ultrashort laser pulse. Journal of Optoeletronics · Laser, 2002, 13(3): 232～236.

[25] Lacourt P A, Dudley J M, Merolla J M, et al. Milliwatt-peak-power pulse characterization at 1.55 um by wavelength-conversion frequency-resolved optical gating. Optics Letters, 2002, 27(10): 863～865.

[26] Barry L P, Delburgo S, Thomsen B C, et al. Optimization of optical data transmitters for 40-Gb/s lightwave systems using frequecy resolved optical gating. IEEE Photon. Tech. Lett., 2002, 14(7): 971～973.

[27] Liu S, Lu D, Zhao L, et al. SHG-FROG characterization of a novel multichannel synchronized AWG-based mode-locked laser//Conference on Lasers and Electro-Optics, OSA Technical Digest (online) (Optical Society of America, 2017), paper JTh2A.131.

[28] Kane D J. Improved principal components generalized projections algorithm for frequency resolved optical gating//Conference on Lasers and Electro-Optics, OSA Technical Digest (online) (Optical Society of America, 2017), paper STu3I.4.

[29] Hyyti J, Escoto E, Steinmeyer G, et al. Interferometric time-domain ptychography for ultrafast pulse characterization. Opt. Lett., 2017, 42: 2185～2188.

[30] Sidorenko, Lahav O, Avnat Z, et al. Ptychographic reconstruction algorithm for frequency-resolved optical gating: Super-resolution and supreme robustness. Optica, 2016, 3: 1320～1330.

[31] Heidt A M, Spangenberg D M, Brügmann M, et al. Improved retrieval of complex supercontinuum pulses from XFROG traces using a ptychographic algorithm. Opt. Lett., 2016, 41: 4903～4906.

[32] Ermolov A, Valtna-Lukner H, Travers J, et al. Characterization of few-fs deep-UV dispersive waves by ultra-broadband transient-grating XFROG. Opt. Lett.,2016, 41: 5535～5538.

[33] Fuji T, Shirai H, Nomura Y. Self-referenced frequency-resolved optical gating capable of carrier-envelope phase determination// Conference on Lasers and Electro-Optics, OSA Technical Digest (2016) (Optical Society of America, 2016), paper SM3I.7.

[34] Steinmeyer A. Interferometric FROG for Ultrafast Spectroscopy on the Few-cycle Scale//Conference on Lasers and Electro-Optics, OSA Technical Digest (2016) (Optical Society of America, 2016), paper STu4I.1.

[35] Okamura B, Sakakibara Y, Omoda E, et al. Experimental analysis of coherent supercontinuum generation and ultrashort pulse generation using cross-correlation frequency resolved optical gating (X-FROG). J. Opt. Soc. Am. B, 2015, 32: 400～406.

[36] Itakura R, Kumada T, Nakano M, et al. Frequency-resolved optical gating for characterization of VUV pulses using ultrafast plasma mirror switching. Opt. Express, 2015, 23: 10914～10924.

[37] Snedden E W, Walsh D A, Jamison S P. Revealing carrier-envelope phase through frequency mixing and interference in frequency resolved optical gating. Opt. Express, 2015, 23: 8507～8518.

[38] Hause A, Kraft S, Rohrmann P, et al. Reliable multiple-pulse reconstruction from second-harmonic-generation frequency-resolved optical gating spectrograms. J. Opt. Soc. Am. B, 2015, 32: 868～877.

[39] Li X J, Liao J L, Nie Y M, et al. Unambiguous demonstration of soliton evolution in slow-light silicon photonic crystal waveguides with SFG-XFROG. Opt. Express, 2015, 23: 10282～10292.

[40] 金永兴,刘涛,康娟, 等.一种新的光纤拉曼增益系数的测量方法. 光通信研究，2008, 148(4): 39～41.

[41] 陈健，张晋，彭江得, 等.光纤拉曼放大器增益系数与噪声系数的实验研究. 中国激光，2001, 28(11): 1021～1023.

[42] Mollenauer L F, Smith K, Gordon J P, et al. Resistance of solitons to the effects of polarization dispersion in optical fibers. Optics Letters, 1989, 14(21): 1219～1221.

[43] Evangelides S G, Mollenauer L F, Gordon J P, et al. Polarization multiplexing with solitons. Journal of Lightwave Technology, 1992, 10(1): 28～35.

[44] Zheng H J, Liu S L, Li X, et al. Temporal characteristics of an optical soliton with distributed Raman amplification. Journal of Applied Physics, 2007, 102(10): 103106-1-4.

[45] 郑宏军, 刘山亮, 田振, 等. 拉曼放大对孤子传输特性的影响研究. 中国激光, 2008, 35(6): 861～866.

[46] 黎昕, 郑宏军, 刘山亮. 利用啁啾脉冲传输测量拉曼增益系数的新方法, 光电子 · 激光, 2011, 22(9): 1395～1400.

第 8 章　新型少模光纤、少模复用器及少模脉冲传输

近年来，云计算、大数据和流媒体等所引发的各种业务流量爆炸式增长，世界各个国家相继制定了国家宽带战略[1]，单模光纤(SMF)通信容限面临趋近香农极限

$$C = W\log_2(1 + S/N) \tag{8-1}$$

的限制。其中，W 为频谱带宽，S/N 是信噪比。光纤通信发展迎来了前所未有的机遇和挑战。光纤通信业界围绕时分(time)复用、波分(frequency，频率)复用、偏振(polarization)复用、正交幅度调制(QAM)、空分(space)复用(包括模分复用、芯式复用和轨道角动量(OAM)复用)五个物理维度对通信网络的传输速率、传输容量进行了不同程度的突破，如图 8-1 所示。其中，最近提出的模分复用是突破通信挑战的最佳物理维度之一[2, 3]。

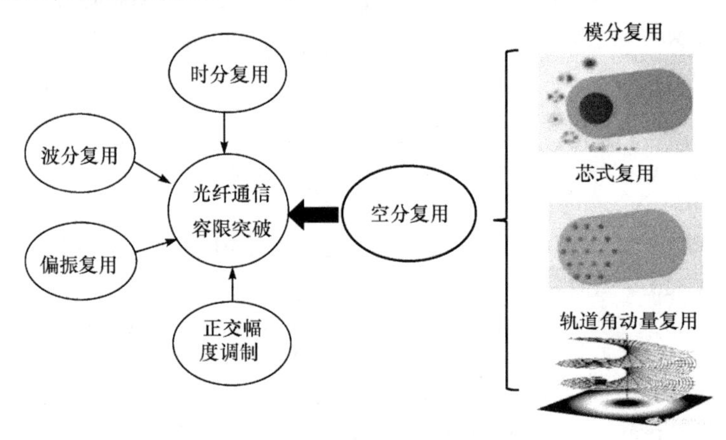

图 8-1　光纤通信容限的突破

在模分复用方面，通常利用光纤每个模式间的正交性，将每个模式作为独立的信道，形成多输入多输出(multiple-input multiple-output，MIMO)通道，提高系统传输容量。其研究主要集中在基于少模光纤(few mode fiber，FMF)的模分复用研究，并且与长距离相干光纤通信相结合，以显著提高通信容量[3]。在不增加功率和带宽条件下，模分复用传输构架每增加一种线偏振模式信号，传输容量比单

模系统的增加一倍，频谱利用率提高一倍，成本有效降低；假设一套单模系统采用 20 个光放大器来实现较远距离传输，要想把系统容量提高到 6 倍，则需要采用 6 套单模系统，需要 120 个光放大器。然而，相同情况下，采用本项目模式复用构架以 6 种模式传输就可以将系统容量提高到 6 倍，因不同模式信号可以共用光放大器，则共需要 20 个光放大器，比单模系统减少 100 个光放大器；系统其他器件也类似地减少，虽然模式复用系统单个器件的复杂度和成本会比单模的高些，但综合考虑下，模式复用系统的成本仍然会大幅度降低。同时少模光纤增大有效面积降低了非线性效应，与多模光纤比较，少模光纤控制了模式数量，优化了模式色散和串扰。

目前，模分复用研究由美国、日本和部分欧洲国家主导[3-31, 33-42]。在中国，2014 年国家科技部资助了天津大学牵头的“多维复用光纤通信基础研究”(“973”基础研究项目[3])。天津大学、清华大学、北京大学、北京邮电大学、中国科学院半导体研究所、武汉邮电科学研究院、华中科技大学、北京交通大学先后加入模分复用领域的研究工作。总之，模分复用领域处于基础研究阶段，已经成为新一代通信系统的研究热点[3-31, 33-42]。模分复用研究涉及模分复用光纤、模式产生与转换、复用解复用机制、模分复用光放大器、模分复用光传输系统、模分复用光网络等；本章主要研究少模光纤、少模复用器、少模光纤系统中的模拟信号传输和少模光纤系统网络中的数字脉冲传输。

8.1　少模光纤和少模复用器

8.1.1　少模光纤的结构与特性

8.1.1.1　折射率阶跃分布单芯少模光纤

图 8-2 所示是折射率阶跃分布单芯少模光纤线性偏振模式图对应光纤横截面积变化情况的示意图[19]。图中，n_1 是纤芯折射率，n_2 是光纤包层折射率，a 是纤基础芯半径，λ 是入射光波长，

$$V=(2\pi a/\lambda)\sqrt{n_1^2-n_2^2} \tag{8-2}$$

是归一化频率，当 V =2.405 时，光纤中出现对应的线性偏振模基模 LP_{01} (即对应单模情况)；在相同入射波长情况下，随着光纤直径的增加，当 2.405<V<3.8 时，光纤中出现对应的线性偏振模基模 LP_{01}、高阶模 LP_{11a} 和 LP_{11b}；其中，LP_{11a} 和 LP_{11b} 折射率近似相同，是简并模式。

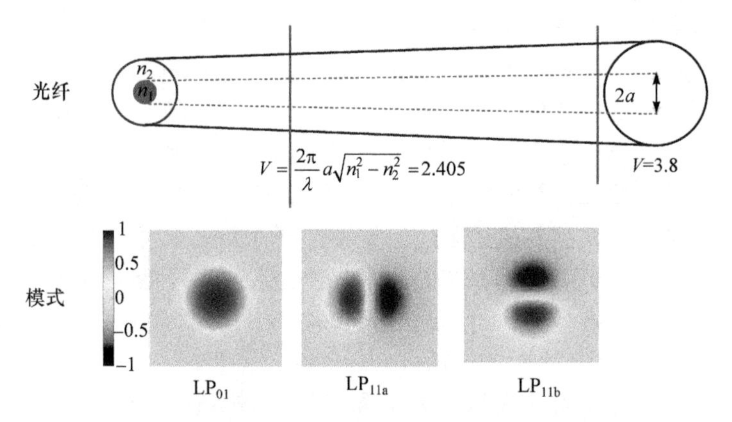

图 8-2　折射率阶跃分布单芯少模光纤线性偏振模式图对应光纤横截面积变化情况的示意图

图 8-3 所示是折射率阶跃分布单芯少模光纤线性偏振(LP)模式归一化传播常数随归一化频率的变化情况[20, 21]。折射率阶跃分布光纤线性偏振模式归一化的传播常数为

$$B=(n_{\mathrm{eff},lm}^2-n_2^2)/(n_1^2-n_2^2) \tag{8-3}$$

式中，$n_{\mathrm{eff},lm}$ 是光纤中线性偏振模式 LP_{lm} 的有效折射率，n_1 是纤芯折射率，n_2 是光纤包层折射率。折射率阶跃分布光纤线性偏振模式归一化传播常数 B 与归一化频率 V 的关系图通常用于光纤设计的参数选择。由图中可见，折射率阶跃分布光纤的各个线性偏振模式归一化传播常数 B 随归一化频率 V 增加而增加；在光纤设计中，通常选取较大的归一化频率 V 数值以期所选模式具有较好的传播

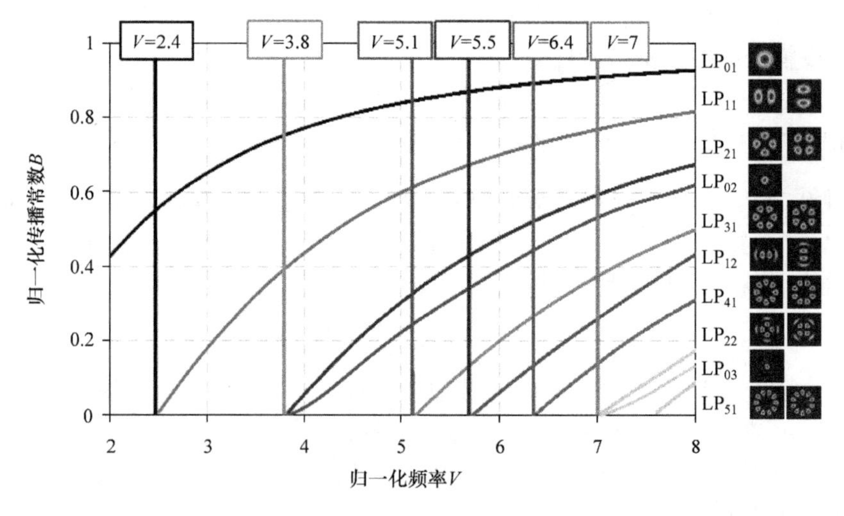

图 8-3　折射率阶跃分布单芯少模光纤线性偏振模式归一化传播常数随归一化频率的变化情况

性能，并避免更高阶模式出现；若要分别设计单模、两模、四模、五模、六模和七模光纤，则归一化频率 V 数值分别选择为 2.4、3.8、5.1、5.5、6.4 和 7；当归一化频率 V 数值选定后，就需要考虑优化纤芯的折射率差 n_1-n_2 的数值；不仅要考虑光纤的弯曲损耗、有效面积随纤芯的折射率差 n_1-n_2 的数值的变化，还需要考虑不同模式之间要有较大的有效折射率差 $|n_{\mathrm{eff},lm}-n_{\mathrm{eff},l'm'}|$，以期减小模式间的耦合、增大差分模式群时延(DMGD)。纤芯的折射率差 n_1-n_2 的数值选择非常复杂，数值太大或太小都不合适，通常情况下需要作折中均衡。一种典型的折射率阶跃分布光纤设计考虑是：在 1550nm 波段，光纤有效折射率差 $|n_{\mathrm{eff},lm}-n_{\mathrm{eff},l'm'}|\geqslant 0.5\times 10^{-3}$，10mm 弯曲半径下的弯曲损耗≤10dB/turn，有效模场面积≥80μm^2。这样的光纤一般具有较大的差分模式群时延(DMGD≥1ns/km)，属于弱耦合少模光纤。

图 8-4 是一种 6-LP 模式单芯少模光纤的折射率阶跃分布示意图[20-22]。由图可见，6-LP 模式光纤的纤芯折射率与标准多模光纤的纤芯中心折射率一致，远大于标准单模光纤的纤芯折射率。6-LP 模式光纤的纤芯半径稍大于标准单模光纤的纤芯半径，远小于标准多模光纤的纤芯半径。

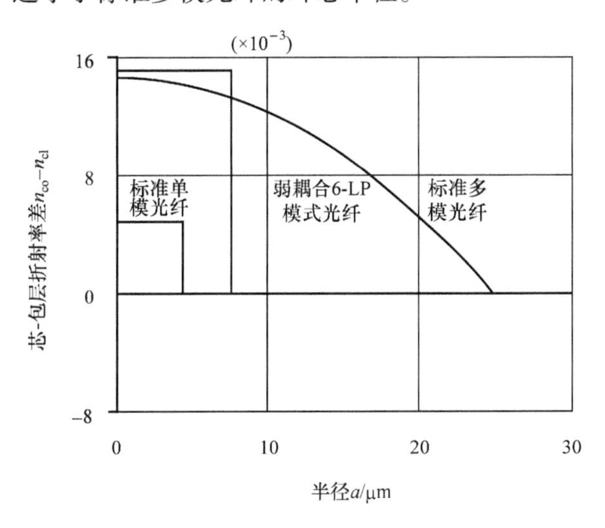

图 8-4　一种 6-LP 模式单芯少模光纤的折射率阶跃分布

表 8-1 给出一种 6-LP 模式折射率阶跃分布单芯少模光纤的参数。由表可见，该光纤的所有六个模式的有效折射率与包层折射率差在 2.43×10^{-3} 以上，弯曲损耗远小于 1dB/turn，不同模式之间的有效折射率差量级为 1×10^{-3}(LP_{21} 与 LP_{02} 的有效折射率差为 1.02×10^{-3})，各个模式的有效模场面积量级为 100μm^2，各个模式与基模之间的差分模式群时延 DMGD 量级为 1ns/km，除模式 LP_{12} 外的各个模式的色散为 20～30ps/(nm · km)(模式 LP_{12} 的色散为−9.6ps/(nm · km))。

表 8-1　一种 6-LP 模式折射率阶跃分布单芯少模光纤的参数

	LP_{01}	LP_{11}	LP_{21}	LP_{02}	LP_{31}	LP_{12}
$n_{eff}-n_{cl}/\times 10^{-3}$	13.48	11.12	8.01	6.99	4.28	2.43
弯曲损耗/(dB/turn)	≪1	≪1	≪1	≪1	≪1	<1
$\mathrm{Min}\lvert\Delta n_{eff}\rvert\times 10^{-3}$			1.02(LP_{21} vs. LP_{02})			
$A_{eff}/\mu m^2$	115	107	113	98	117	112
DMGD vs. LP_{01}/(ns/km)	—	5.56	1.45	1.35	1.89	1.59
色散/[ps/(nm · km)]	22.4	26.9	29.9	27.1	28.2	−9.6

8.1.1.2　折射率渐变分布单芯少模光纤

图 8-5 所示是折射率渐变分布单芯少模光纤线性偏振模式归一化传播常数随归一化频率的变化情况[21, 22]。从图中可见，折射率渐变分布单芯少模光纤的各个线性偏振模式归一化传播常数 B 随归一化频率 V 增加而增加，该变化率比折射率阶跃分布光纤的要小。在渐变光纤设计中，选取归一化频率 V 数值的原则与折射率阶跃分布光纤的参数选择一致；若要分别设计单模、两模、四模和六模光纤，则归一化频率 V 数值分别选择为 3.7、5.7、7.8 和 9.8。当归一化频率 V 数值选定后，就需要考虑优化纤芯的折射率差 $n_1 - n_2$ 的数值；一种典型的折射率渐变分布光纤设计考虑是：在 1550nm 波段，10mm 弯曲半径下的弯曲损耗≤10dB/turn，这样的光纤一般具有较小的差分模式群时延(当 LP 模式多于 4 时，DMGD≥0.1ns/km)，属于强耦合少模光纤。

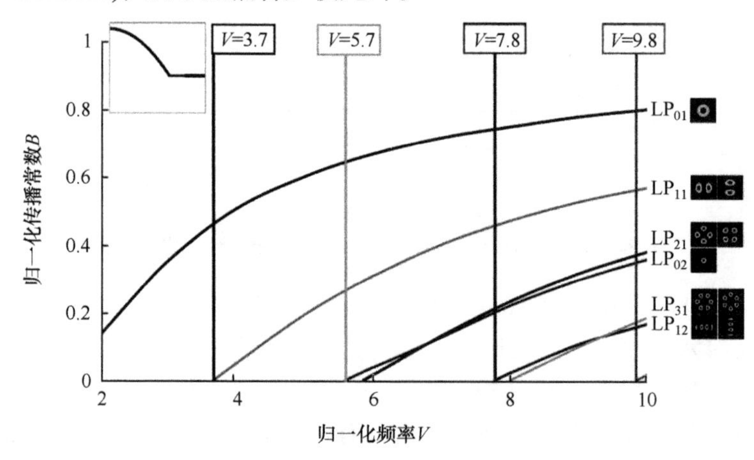

图 8-5　折射率渐变分布单芯少模光纤线性偏振模式归一化传播常数随归一化频率的变化情况

图 8-6 所示是一种 6-LP 模式单芯少模光纤的折射率渐变分布示意图[21, 22]。

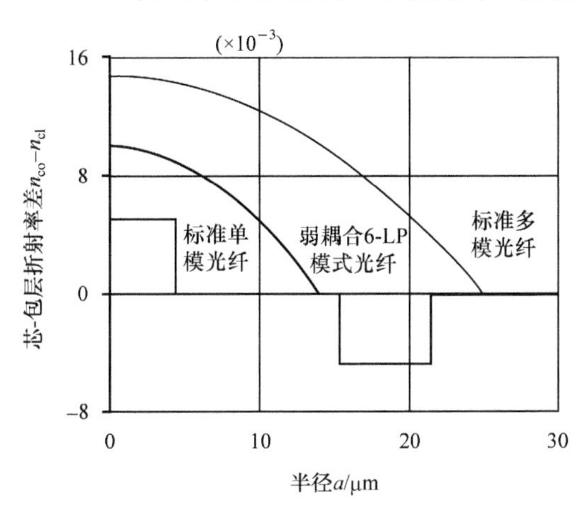

图 8-6　一种 6-LP 模式单芯少模光纤的折射率阶跃分布

从图 8-6 中可见，6-LP 模式渐变光纤的纤芯折射率在标准多模光纤的纤芯中心折射率和标准单模光纤的纤芯折射率之间；纤芯中心到包层的折射率呈渐变型分布，折射率按照方程[19]

$$n(r)=\begin{cases} n_1\cdot[1-2\Delta(r/a)^{\alpha}]^{1/2}, & r\leqslant a \\ n_2, & r>a \end{cases} \tag{8-4}$$

设置，式中 n_1 表示纤芯中心折射率，n_2 表示包层折射率，r 表示光纤中任意一点到轴心的距离，a=14μm 是光纤纤芯外径，α=1.95，n_1-n_2=10×10^{-3}，$\Delta=(n_1^2-n_2^2)/(2n_1^2)\approx(n_1-n_2)/n_1$，$V$=9.65(1550nm)。为了减小弯曲损耗，该光纤纤芯外部加了一个折射率下陷层。6-LP 模式渐变光纤的纤芯半径大于标准单模光纤的纤芯半径，小于标准多模光纤的纤芯半径。

表 8-2 给出一种 6-LP 模式折射率渐变分布单芯少模光纤的参数。由表可见，该光纤的所有六个模式的有效折射率与包层折射率差在 1.46×10^{-3} 以上，弯曲损耗远小于 1dB/turn，各个模式与基模之间的差分模式群时延 DMGD 量级为 −6ps/km～−8ps/km(模式 LP_{31} 的 DMGD 是 0.8ps/km)，各个模式的有效模场面积量级为 100～300μm^2，各个模式的色散近似为 20ps/(nm·km)。

表 8-2　一种 6-LP 模式折射率渐变分布单芯少模光纤的参数

	LP_{01}	LP_{11}	LP_{21}	LP_{02}	LP_{31}	LP_{12}
$n_{eff}-n_{cl}$/×10^{-3}	7.71	5.62	3.54	3.55	1.46	1.48
弯曲损耗/(dB/turn)	≪1	≪1	≪1	≪1	<10	<10
DMGD vs. LP_{01}/(ps/km)	—	−7.6	−6.5	−7.8	0.8	−7.5

续表

	LP_{01}	LP_{11}	LP_{21}	LP_{02}	LP_{31}	LP_{12}
Max|DMGD|/(ps/km)	8.6(LP_{31} vs. LP_{02})					
A_{eff}/μm^2	126	169	227	256	273	274
色散/[ps/(nm · km)]	20.1	20.4	20.6	20.6	20.9	20.9

8.1.1.3 多芯少模光纤的特性

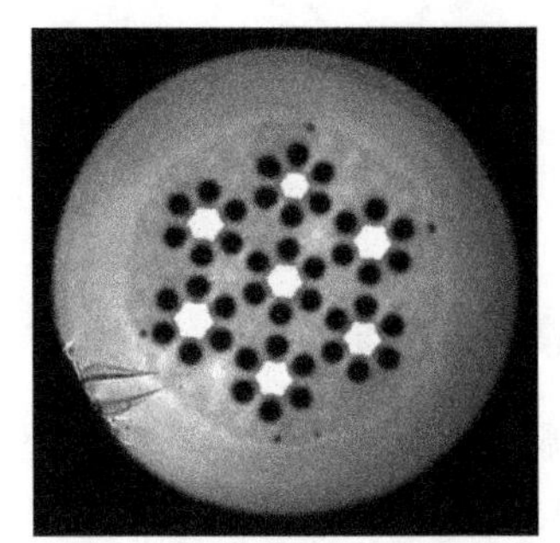

图 8-7 一种周边空气孔辅助型七芯少模光纤的横截面

图 8-7 所示是一种周边空气孔辅助型七芯少模光纤的横截面[23]。图中，七个亮斑分别是该光纤的七个纤芯；每个纤芯周边的六个黑色斑点是六个空气孔。

表 8-3 是一种周边空气孔辅助型七芯少模光纤的设计参数。由表可见，该光纤的纤芯与包层的折射率差为 0.36%，纤芯直径为 13.1μm，纤芯距为 40μm，空气孔直径为 8.2μm，空气孔间距为 13.3μm，包层直径为 192μm，涂覆层直径为 375μm。

表 8-3 一种周边空气孔辅助型七芯少模光纤的设计参数

参数	数值
芯-包层折射率差	0.36%
纤芯直径	13.1μm
纤芯距	40μm
空气孔直径	8.2μm
空气孔间距	13.3μm
空气孔径与间距比	0.62
包层直径	192μm
涂覆层直径	375μm

表 8-4 给出一种周边空气孔辅助型多芯少模光纤的实验测量和仿真参数。模式 LP_{01} 的模场直径实验测量值为 11.8μm；模式 LP_{01} 和 LP_{11} 的有效面积仿真值分别为 113μm^2 和 170μm^2；模式 LP_{11} 的截止波长仿真值为 2100nm；模式 LP_{01} 和 LP_{11} 的色散仿真值分别为 23ps/(nm · km)和 28ps/(nm · km)，色散斜率仿真值分别

为 0.06ps/(nm^2 · km)和 0.07ps/(nm^2 · km)；有效折射率差仿真值为 2.4×10^{-3}；差分模式群时延实验测量值为 4.6ps/m；模式 LP_{01} 和 LP_{11} 的芯间串扰仿真值分别为 −100dB/km 和−70dB/km，实验测量值分别为−60dB/km 和− 40dB/km。还可以设计更多纤芯、更多模式的多芯少模光纤以实现更高通信容量。

表 8-4 一种周边空气孔辅助型多芯少模光纤的实验测量和仿真参数

光纤参数	LP_{01}	LP_{11}
模场直径实验测量值(MFD)/μm	11.8	—
有效面积仿真值(A_{eff})/μm^2	113	170
截止波长仿真值(λ_c)/nm	—	~ 2100
色散仿真值/[ps/(nm · km)]	23	28
色散斜率仿真值/[ps/(nm^2 · km)]	0.06	0.07
有效折射率差仿真值(ΔN_{eff})	2.4×10^{-3}	
差分模式群时延实验测量值(DGD)/(ps/m)	4.6	
芯间串扰仿真值/(dB/km)	−100	−70

8.1.2 少模模分复用器的结构与特性

少模模分复用器是将多种模式信号复用到同一个信道进行传输的器件，是实现模分复用系统的重要组成部分。

8.1.2.1 基于相位板的模分复用器

图 8-8 所示是一种基于相位板的三模式模分复用器[24]，其中(a)为三模式模分复用器结构图，(b)为理论计算的各模式光强模斑，(c)为实验测量的各模式光强模斑。由结构图可得，该基于相位板的模分复用器，实现了基模到高阶模式的转换，并将三种模式信号复用到同一个三模式光纤信道。由各模式光强模斑图可得，理论计算得到的各模式光强模斑与实验测量结果一致。该模分复用器的模式 LP_{01}、LP_{11a} 和 LP_{11b} 的耦合损耗分别为 8.3dB、10.6dB 和 9.0dB。这种模分复用器要求空间耦合校准具有高精确度，自由空间耦合的耦合损耗大且难于集成，成本高。

图 8-9 所示各个模式的光强模斑与相位板的关系图[24]。由图 8-9 可得，若同时考虑 *XY* 偏振复用，三模式模分复用器可以实现六种模式复用；其中基模可以不要相位板，直接入射即可；模式 LP_{11} 的四个简并模式可以通过相位差为 π 的不同全息相位板来形成。

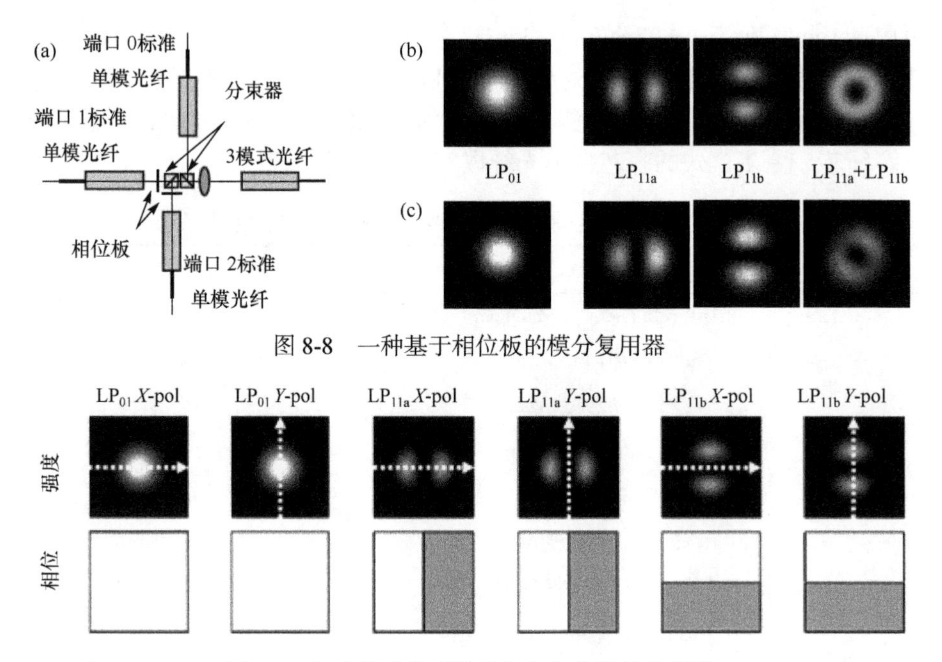

图 8-8　一种基于相位板的模分复用器

图 8-9　各个模式的光强模斑与相位板的关系图

8.1.2.2　基于液晶空间光调制器的模分复用器

图 8-10 所示是基于液晶空间光调制器(LCOS-based SLM)的模分复用器[24]，其中(a)为该模分复用器结构图，(b)为实验测量的模式光强模斑，(c)为实验测量的模式干涉模斑。图中 LCOS 是指基于硅基的液晶，可以受到精确的电压控制，并能调制光束的波前，从而实现入射光的模式转换和复用。

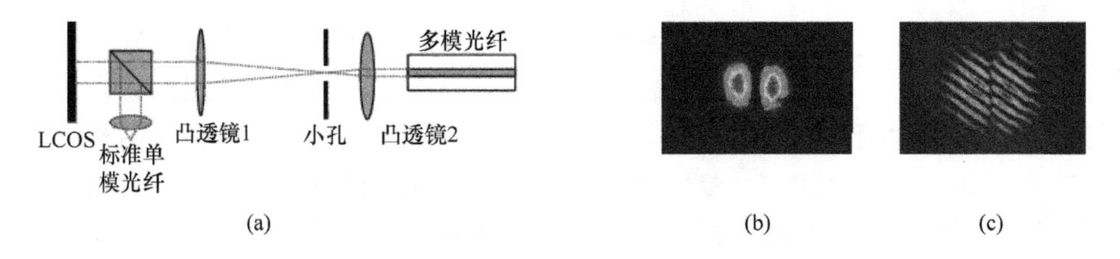

图 8-10　基于液晶空间光调制器的模分复用器

8.1.2.3　光子灯笼

图 8-11 所示是对称标准型三输入端的光子灯笼[19]，其中(a)为该光子灯笼结构图，(b)为光纤拉锥变换区形成的超模模式图，(c)为光纤拉锥少模区形成的少模模式图。

由图 8-11 可见，该光子灯笼是一种对称型三输入端的拉锥形状的模分复用器，实现了三路单模信号转换模式并复用到同一个少模信道。光子灯笼的结构复杂程度虽然比基于相位板或液晶空间调制器的模分复用器低，但是其制造工艺精密、复杂，精确地控制光纤熔融时的位置、拉锥体粗细和长度具有很大难度；通过拉锥形成的输出端口与传输通道不完全匹配，会导致较高损耗。

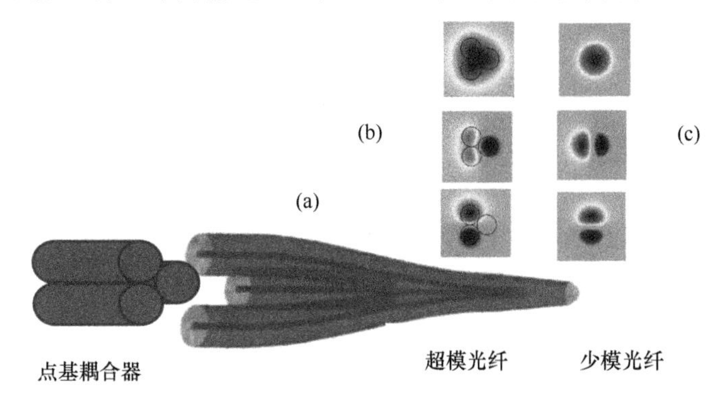

图 8-11　对称标准型三输入端的标准光子灯笼

光子灯笼可分为对称标准型[19, 25, 26]、模式选择型[27, 28]和模组选择型[29-31]。对称标准型光子灯笼[25, 26]，输入光纤的芯径和类型相同，需要与偏振控制器联合控制所产生的模式；后来发展出的模式可选择光子灯笼[27, 28]以及模式组可选择光子灯笼[29-31]，输入光纤的芯径不同，可对产生的模式或模式组作出选择分类；这些模分复用器/解复用器可以是光纤光子灯笼[25-30]，也可以是基于激光刻写 3D 波导的光子灯笼[26, 31]，可以广泛应用于模分复用光通信领域。

8.2　少模光纤系统中的模拟信号传输

近年来，模拟传输技术已经成为实现宽带接入的研究热点之一，应用前景广阔。与传统光纤通信相比，模拟传输系统对光器件的性能以及系统色散、非线性等的要求更为苛刻，导致国内外模拟传输系统的传输距离一般较小。然而，考虑到我国地理环境复杂性和现代信息社会通信需求的迅猛增长，实现无中继、长距离光纤模拟传输变得日益迫切。高增益、高信噪比和大无杂散动态范围(SFDR)的模拟光纤传输链路能够增加传输距离、提高接入网络的分光比和减少用户使用成本[32, 33]，是实现无中继、长距离模拟光纤传输网络的重要途径。然而，在光纤非线性效应作用下，高增益、高信噪比和大无杂散动态范围的模拟光纤传输链路受限于模拟信号光功率。首先，尽管信号输入光功率增加，但光纤中后向受激布里渊散射效应会导致接收端信号的光功率受限；其次，信号高输入光功率引起的克尔非线性效应会导致信号交调畸变。鉴于此，我们率先提出了一种少模光纤模拟传输链路来克服上述讨论中的非线性效应[34]。该实验传输链路的理论依据是，少模光纤的有效模场面积大于单模标准光纤的有效模场面积；高阶模式与基模的串扰可以通过光纤设计参数优化来抑制掉。在我们的实验演示中，模拟信号加载到少模光纤中的基模模式上传输 76km，其传输特性与相同长度单模标准光纤中的特性进行了比较。结果表明，模拟信号在少模光纤中的传输特性优于在单模标准光纤中的特性；与单模标准光纤相比，在少模光纤中的 SFDR 提高了 1.5dB。

8.2.1　少模光纤系统中的模拟传输实验装置

图 8-12 所示是少模光纤系统中的模拟传输实验装置[34]；图中，DP-MZM 是双并行马赫-曾德尔调制器，OA 是光放大器，OBPF 是带通滤波器，OSA 是光谱分析仪，PD 是光探测器。此外还有 1550nm 波段激光器，电模拟信号产生器(signal generator)，电移相器(phase shift)。电模拟信号产生器、电移相器产生两通道电信号，激光器产生 1550nm 的激光；在双并行马赫-曾德尔调制器 DP-MZM 中，两通道电信号调制 1550nm 激光产生两通道调制光波；该调制光波被放大、滤波后分别输入单模标准光纤和少模光纤进行传输。在 1550nm 波段，少模光纤支持两个模式 LP_{01} 和 LP_{11}，模拟传输仅采用基模 LP01，纤芯半径为 6μm，基模 LP_{01} 的有效面积、衰减和色散分别为 $130\mu m^2$、0.201dB/km 和 20.5ps/(km · nm)；标准单模光纤的纤芯半径、衰减和色散分别为 4μm、0.193dB/km 和 17.7ps/(km · nm)；少模光纤基模 LP_{01} 的有效面积比单模标准光纤的要大 65%，这样导致少模光纤中输入光功率较大，但非线性却较小。模拟信号调制格式采用载波和边带近似相等的特殊单边带调制(SSB)；与传统的双边带调制(DSB)格式比较，该单边带调制可以有效减小色散影响和提高受激布里渊的散射阈值。

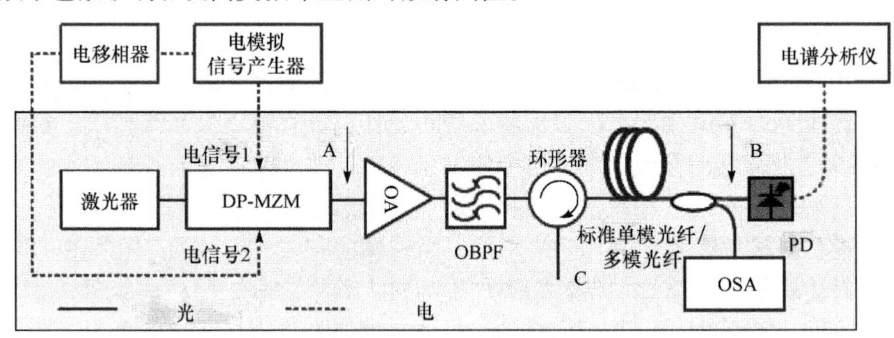

图 8-12　少模光纤系统中的模拟传输实验装置

8.2.2 实验结果与讨论

图 8-13 所示是少模光纤系统中输出功率、后向散射功率与输入光功率的变化关系。图中可见，在少模光纤中，SBS 阈值功率近似达到 14dBm，比标准单模光纤中的 SBS 阈值功率大 3dB。这导致少模光纤中的线性区最大输入光功率达到 15dBm，从而使得在输出端有较高的 RF 电功率。需要注意的是，标准单模光纤的衰减比少模光纤的损耗小 1 dB，导致在线性区有较高的输出功率。

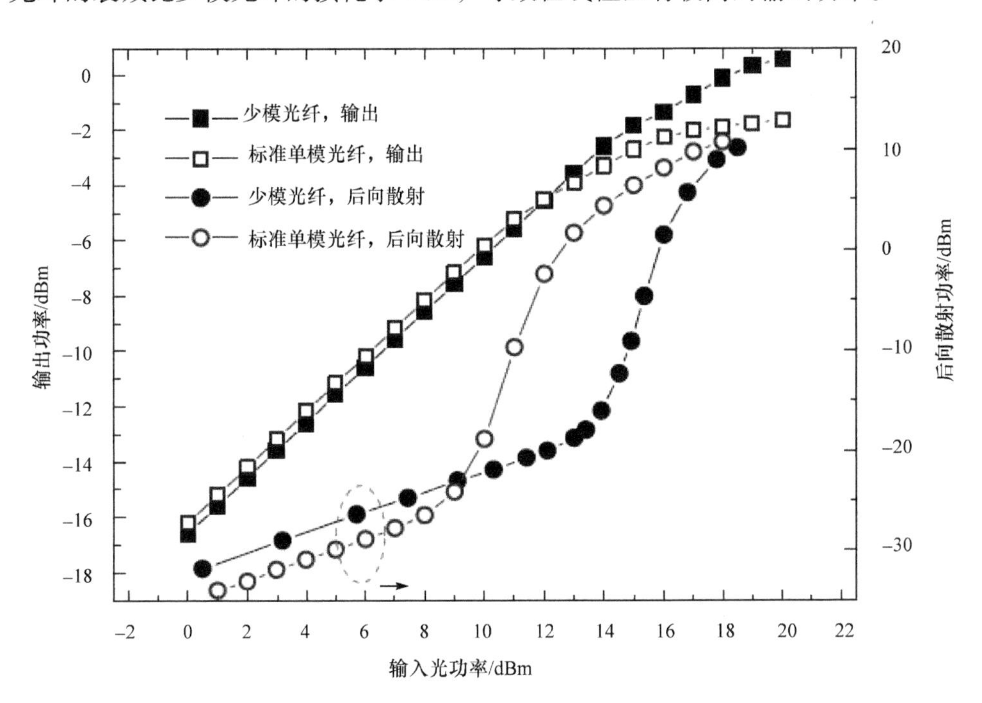

图 8-13 少模光纤系统中输出功率、后向散射功率与输入光功率的变化关系

图 8-14 所示是在 SMF、FMF 中传输前后的光谱，光功率均为 10.6dBm。从图中可见，传输前背靠背的光谱中看不到二级边带；信号传输后，由于光纤克尔非线性诱导的四波混频效应作用，标准单模光纤中光谱出现明显的二级边带，比少模光纤中的二级边带分别大 2.6dB 和 3.3dB；这表明在 SBS 阈值功率以下，标准单模光纤中的克尔非线性比少模光纤中非线性要严重得多。同时，与传统的 SBB 调制格式比较，载波和边带近似相等的特殊 SBB 调制格式中载波功率大幅度减小，使得 SBS 阈值功率有效增加。

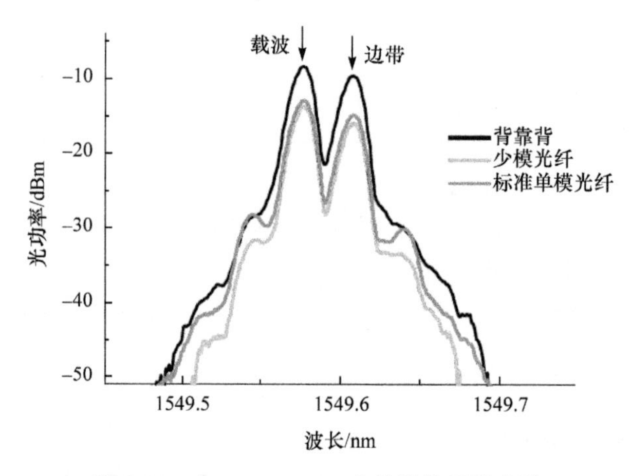

图 8-14　在 SMF、FMF 中传输前后的光谱

图 8-15 所示是输出基波和三阶交调(IMD3)的 RF 电功率与输入光功率的变化关系。在 SBS 阈值功率下的给定光功率，由于标准单模光纤的衰减较小，在标准单模光纤中输出的基波 RF 功率比相应的少模光纤中功率大 2dB；同时，在标准单模光纤中输出的三阶交调功率比相应的少模光纤中功率大 5dB。这是光纤非线性把载波功率转变到三阶交调功率的缘故。当输入光功率增加达到 SBS 作用区域，输出基波和三阶交调功率都停止线性增加；由于少模光纤有较高的 SBS 阈值，尽管少模光纤比单模光纤衰减大，但少模光纤中输出基波和三阶交调功率都大于标准单模光纤中的功率。这样，模拟少模光纤传输链路就可以采用较大的输入光功率以实现较长跨距的传输。

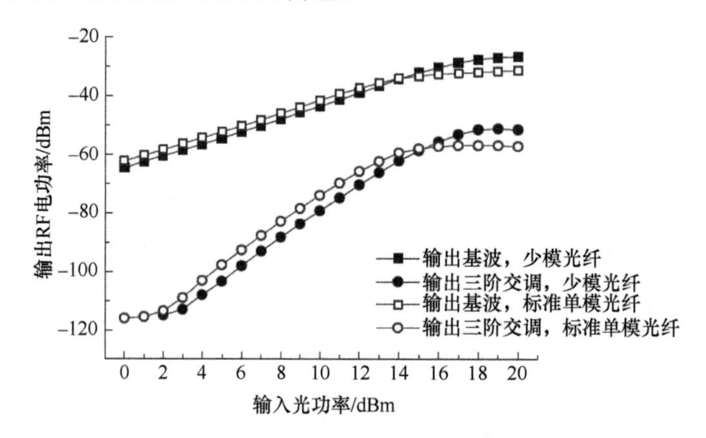

图 8-15　输出 RF 电功率与输入光功率的变化关系

图 8-16 所示是掺铒光纤放大器设置为 33dB 增益模式时输出 RF 电功率与输入 RF 电功率的变化关系。由图可知，在 SBS 和克尔非线性作用下，少模光纤输出端中的基波功率大于标准单模光纤中基波功率 1dB，少模光纤输出端中的三阶交调功率小于标准单模光纤中功率 1dB。这导致了少模光纤中无杂散动态范围比标准单模光纤中的大 1.5dB，这正是少模光纤可以采用较高的输入光功率和能够减小光纤非线性的缘故。

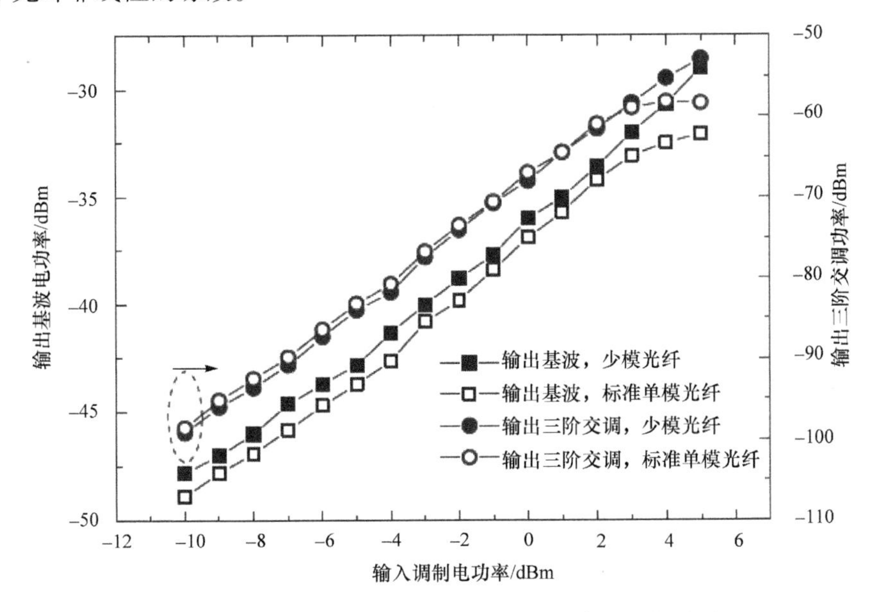

图 8-16　输出 RF 电功率与输入 RF 电功率的变化关系

综上所述，本节首次实验演示了基于少模光纤的基模长跨距模拟传输光纤链路系统，比单模光纤链路提高 3dB 增益，无杂散动态范围提高 1.5dB。若采用更大有效面积少模光纤的高阶模式模拟传输光纤链路系统，则无杂散动态范围提高到 8dB；研究光纤高阶模式波分复用器传输通道串扰，使用高阶模式比常规单模波分复用器传输减小 15dB 串扰[35]。

8.3　少模光纤系统网络中的数字脉冲传输

在接入网中，功率预算和成本是两个非常重要的问题。一个很大的挑战是最大限度地减少上行信道中合波功率损耗以提高功率预算。空间模式复用提供了最大限度地减少上行信道中合波功率损耗却没有显著增加成本的可能性[36]。最近，为了减小上行信道中合波功率的损耗，一种少模(FM)时分复用(TDM)无源光网络

(PON)系统[37, 38]通过使用一种光子灯笼(PL)来取代传统单模合波器。在这项研究中，尽管 FM-PON 有较大的理论优势，但总功率预算的净增加却没有实现，主要是因为：①光子灯笼本身的器件损耗，②光子灯笼与少模光纤的对接耦合损耗，③空间模式少、仅使用三个空间模式。为了提高功率预算的净增加以及满足实际无源光系统需求，要求较大的合波比、更多的模式数目，而这是采用模分复用技术可以实现的；若进一步结合时分/波分复用，能够进一步增加容量、用户数量和传输距离[39, 40]。

结合商业的吉比特无源光网络(GPON)，我们提出了第一个六模式无源光网络，实现了 4dB 的功率预算净增益[41]。这是第一个使用光子灯笼与少模光纤熔接的少模无源光网络。把少模光纤实际用于无源光网络系统方面，这是一个重大成就。与以前的研究比较[42]，我们建立了 6 个模式信道的双向连接，包括一个新增加的模式 LP_{02} 信道，改进了到达无源光网络收发器的功率耦合方式。

8.3.1　少模光纤系统网络中的数字传输实验装置

图 8-17 所示为少模光纤系统网络中的数字传输实验装置[41]。其中，(a)为系统总体连接图，(b)为原有的无源光网络收发器前端，(c)为改进的无源光网络收发器前端，(d)为建议的无源光网络收发器前端。图(a)中 20km 少模光纤和一个模式可选择的光子灯笼被集成到一个商用的 GPON 系统网络中；GPON 包括一个华为的光线路终端(OLT)和八个光网络单元(ONU)；少模光纤六种模式分别支持不同的光网络单元，一个模式至少支持一个光网络单元；实际应用中，不同的光网络单元可以分组到不同的少模模式。例如，若有 64 个光网络单元，则 10 或 11 个光网络单元可以作为一个分组共同连接到同一个少模模式。图(b)中原有的无源光网络收发器前端中有一段单模光纤，会导致高阶模式有较高的损耗，特别是模式 LP_{02} 的损耗最大，以至于很难建立该模式信道的 GPON 传输链接。图(c)中我们改进了无源光网络收发器前端，通过一个微球型透镜把少模光纤的各个模式耦合到光电探测器(PD)，解决了高阶模式耦合损耗大的问题。图(d)是我们建议的另一种无源光网络收发器前端，可以通过一段多模光纤把少模光纤的各个模式耦合到光电探测器，也能解决高阶模式耦合损耗大的问题。

图 8-18 为少模光纤系统网络中光子灯笼横截面及其模式图[41]。其中，(a)是光子灯笼的横截面，(b)是光子灯笼各个模式图；该光子灯笼是采用不同芯径的输入光纤和低折射率的包层管拉锥而成的模式可选择的光子灯笼。从模式图可以看出，光子灯笼性能很好，光子灯笼本身损耗小于 1dB。

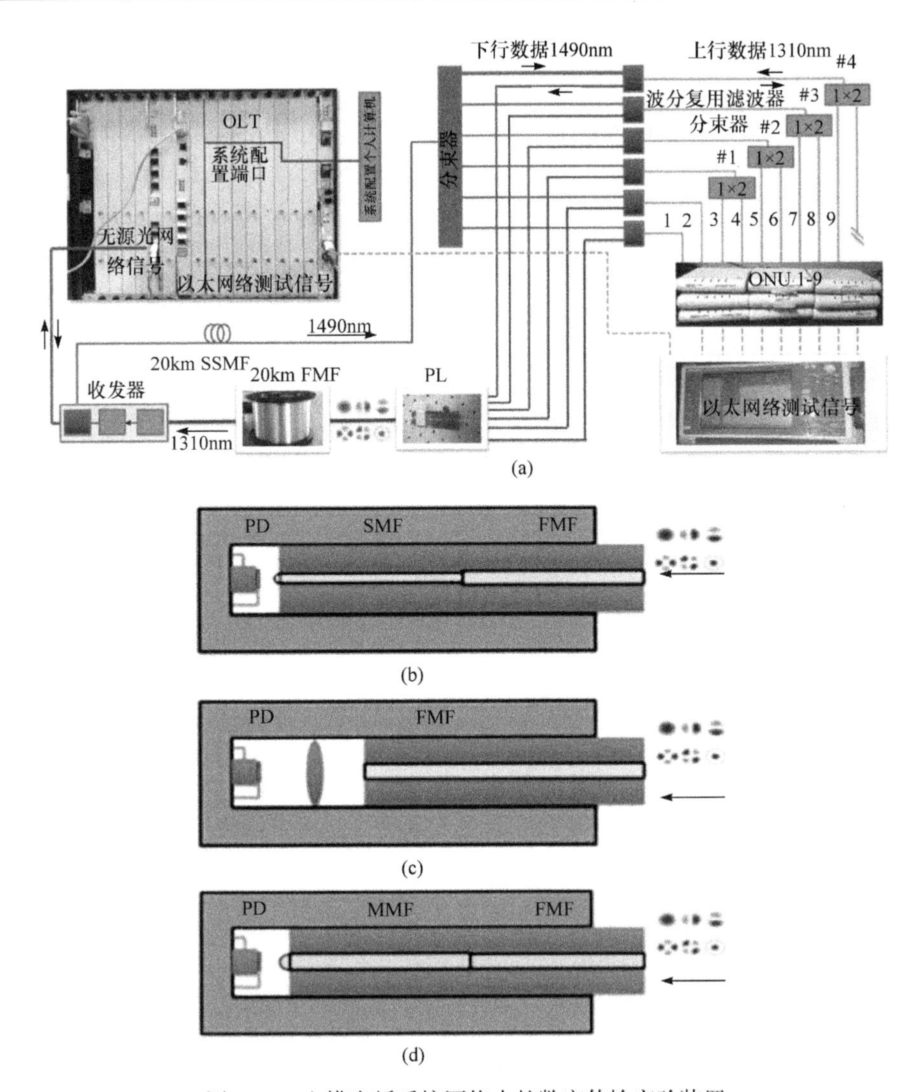

图 8-17　少模光纤系统网络中的数字传输实验装置

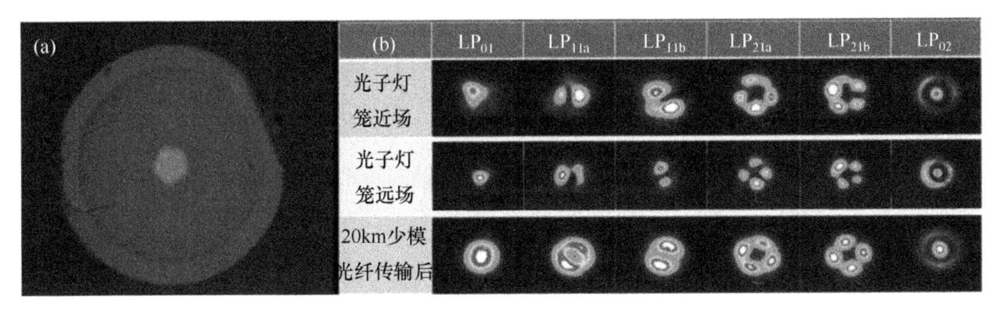

图 8-18　少模光纤系统网络中光子灯笼的横截面及其模式图

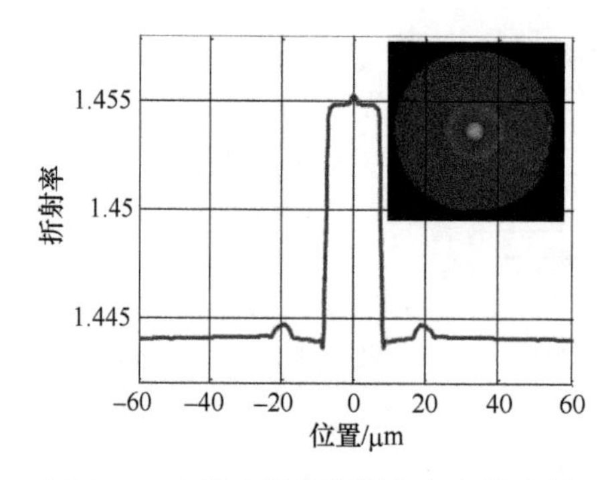

图 8-19 少模光纤系统网络中少模光纤的横截面及其纤芯折射率分布

图 8-19 为少模光纤系统网络中少模光纤的横截面及其纤芯折射率分布[41]；在 1300nm 波段，该少模光纤 20km 与光子灯笼熔接后各模式 LP_{01}、LP_{11a}、LP_{11b}、LP_{21a} 和 LP_{21b} 的总损耗分别为 10.1dB、12.5dB、9.5dB、10.1dB 和 10.1dB。少模光纤各个模式的衰减约为 0.4dB/km；本系统总体损耗比采用常规合波器的情况要少 4dB。在 1300/1550nm 波段，少模光纤各个模式组的色散分别为 1.5/21ps/(nm · km)、4.5/26ps/(nm · km)、6.5/19ps/(nm · km)和 4/8ps/(nm · km)。

图 8-20 为少模光纤各个模式在 1310nm 的脉冲响应[41]。图中可见，各个模式相对于模式 LP_{01} 都存在不同的延迟，模式串扰较小；模式 LP_{11a} 和 LP_{21a} 的串扰比其他模式的稍大些。

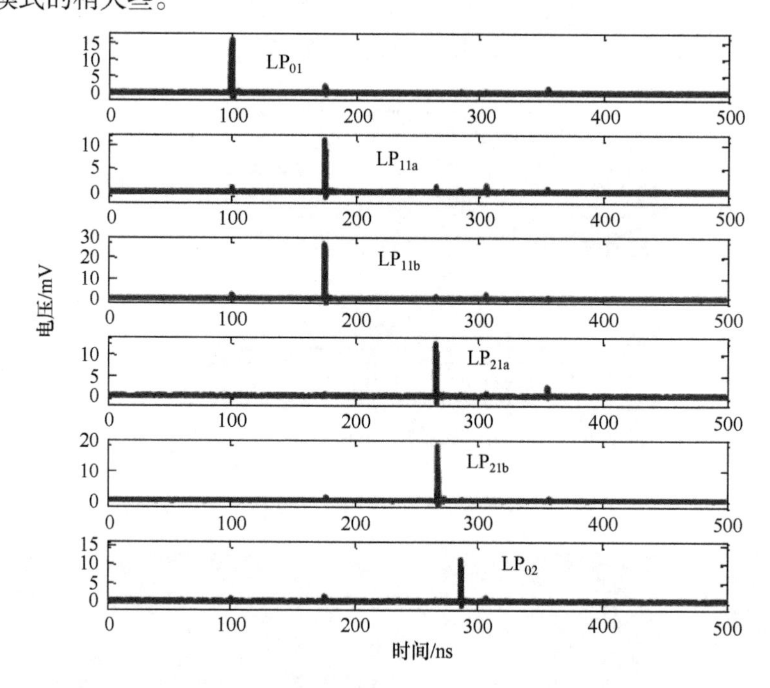

图 8-20 少模光纤各个模式在 1310 nm 的脉冲响应

8.3.2 少模数字脉冲传输特性的实验测量与分析

图 8-21 为少模光纤系统网络中各个模式的传输眼图[41]；由图可见，各个模

式的眼图较清晰，但都存在一定过冲和较厚眼皮，表明存在一定的模式串扰。

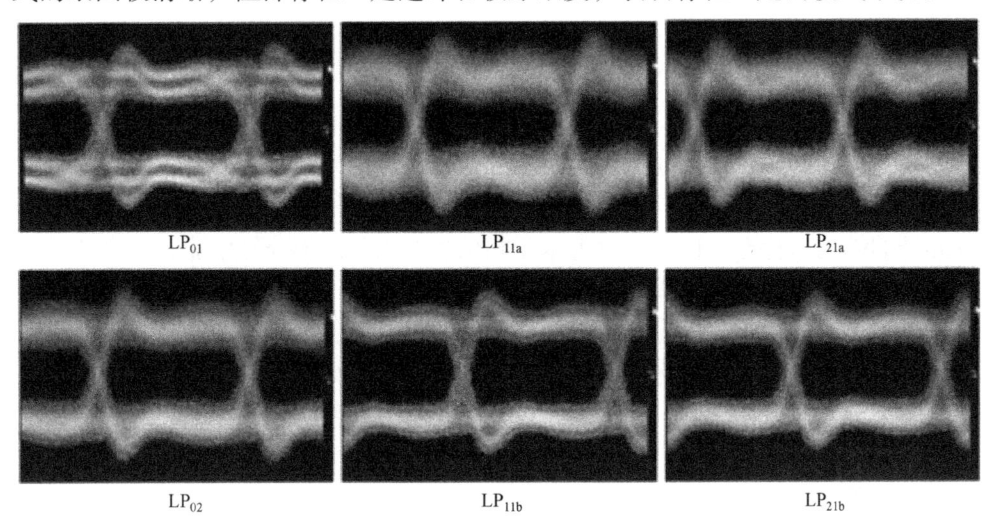

图 8-21　少模光纤系统网络中各个模式的传输眼图

图 8-22 是少模光纤系统网络中各个模式误码率随接收光功率的变化关系；各个模式的误码率(BER)均能达到 10^{-9} 以下；从模式 LP_{11a} 和 LP_{21a} 的误码率曲线的斜率可以看到，这两个模式的串扰要大些，这与两个模式的脉冲响应对应。

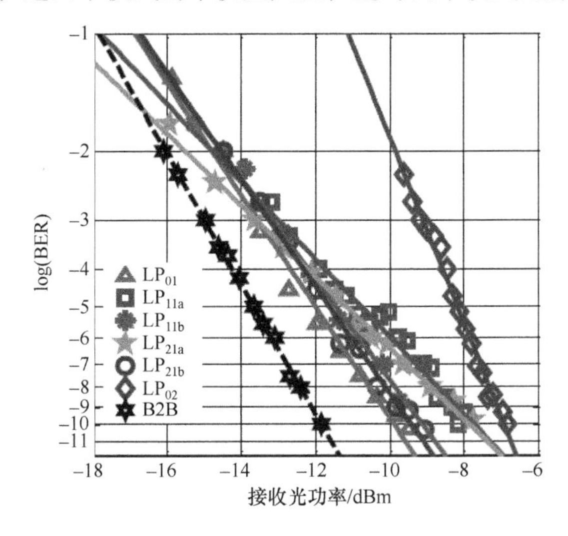

图 8-22　少模光纤系统网络中各个模式误码率随接收光功率的变化关系

综上所述，在少模模式数字传输方面，采用六模式光子灯笼和六模式少模光纤，在上行 PON 中提高了 4dB 功率预算，各个模式的误码率均低于 10^{-9}。该工作为少模模式复用解复用及其模式数字传输的进一步研究奠定了研究基础。

8.4 本章小结

本章介绍了一种六模式折射率阶跃分布单芯少模光纤的结构与特性、一种六模式折射率渐变分布单芯少模光纤的结构与特性、一种周边空气孔辅助型七芯少模光纤的结构与特性；讨论了一种基于相位板的三模式模分复用器、基于液晶空间光调制器的模分复用器以及光子灯笼的结构与特性；在少模模式模拟传输方面，率先提出了一种基于少模光纤的模拟传输链路；结果表明，模拟信号在少模光纤中的传输特性优于在单模标准光纤中的特性；与单模标准光纤相比，在少模光纤中的无杂散动态范围提高了 1.5dB；若采用更大有效面积的少模光纤的高阶模式模拟传输光纤链路系统，则 SFDR 提高更多[34, 35]。在少模模式数字传输方面，首次提出了基于光子灯笼与少模光纤熔的六模式无源光网络，在上行无源光网络中实现了 4dB 的功率预算净增益，各个模式的误码率均优于 10^{-9}[41, 42]。本章工作为少模模分复用、解复用及其模式传输的进一步研究奠定了基础。

参考文献

[1] Kalil T (Obama Administration， Deputy Director for Policy at OSTP). Big data research and development initiative (2012)(http://www.whitehouse.gov/blog /2012/03/29/big-data-big-deal).

[2] Wang J, Yang J Y, Fazal I M, et al. Terabit free-space data transmission employing orbital angular momentum multiplexing. Nature Photonics, 2012, 6(7): 488～496.

[3] Richardson D J, Fini J M, Nelson L E. Space-division multiplexing in optical fibes. Nature Photonics, 2013, 7: 354～362.

[4] Ryf R, Mestre M A, Randel S, et al. Combined SDM and WDM transmission over 700-km few-mode fiber. Optical Fiber Communication Conference, Optical Society of America, 2013.

[5] Mori T, Sakamoto T, Wada M, et al. Few-mode fibers supporting more than two LP modes for mode-division-multiplexed transmission with MIMO DSP. Journal of Lightwave Technology, 2014, 32(14): 2468～2479.

[6] Lim E L, Jung Y, Kang Q, et al. First demonstration of cladding pumped few-moded EDFA for mode division multiplexed transmission//Optical Fiber Communication Conference, OSA Technical Digest (Optical Society of America, 2014), paper M2J.2.

[7] http://modegap.eu/.

[8] Tobita Y, Fujisawa T, Sakamoto T, et al. Optimal design of 4LP-mode multicore fibers for high spatial multiplicity. Opt. Express, 2017, 25: 5697～5709.

[9] Parmigiani F, Jung Y, Horak P, et al. C- to L-band wavelength conversion enabled by parametric processes in a few mode fiber//Optical Fiber Communication Conference, OSA Technical Digest (online) (Optical Society of America, 2017), paper Th1F.4.

[10] Du J, Xie D Q, Yang C, et al. Demonstration of analog links using spatial modes in km-scale few mode fiber. Opt. Express, 2017, 25: 3613～3620.

[11] Soma D, Wakayama Y, Igarashi K, et al. Partial MIMO-based 10-mode-multiplexed transmission over 81km weakly-coupled few-mode fiber//Optical Fiber Communication Conference, OSA Technical Digest (online) (Optical Society of America, 2017), paper M2D.4.

[12] Li J X, Du J B, Ma L, Li M J, et al. Second-order few-mode Raman amplifier for mode-division multiplexed optical communication systems. Opt. Express, 2017, 25: 810～820.

[13] Ferreira F, Sanchez C, Suibhne N, et al. Nonlinear transmission performance in delay-managed few-mode fiber links with intermediate coupling//Optical Fiber Communication Conference, OSA Technical Digest (online) (Optical Society of America, 2017), paper Th2A.53.

[14] Chang S H, Moon S R, Chen H S, et al. All-fiber 6-mode multiplexers based on fiber mode selective couplers. Opt. Express, 2017, 25: 5734～5741.

[15] Melati D, Alippi A, Annoni A, et al. Integrated all-optical MIMO demultiplexer for mode- and wavelength-division-multiplexed transmission. Opt. Lett., 2017, 42: 342～345.

[16] Liu H, Wen H, Correa R A, et al. Reducing group delay spread in a 9-LP mode FMF using uniform long-period gratings//Optical Fiber Communication Conference, OSA Technical Digest (online) (Optical Society of America, 2017), paper Tu2J.5.

[17] Liu H, Wen H, Zacarias J C A, et al. 3×10 Gb/s mode group-multiplexed transmission over a 20 km few-mode fiber using photonic lanterns//Optical Fiber Communication Conference, OSA Technical Digest (online) (Optical Society of America, 2017), paper M2D.5.

[18] Sillard P, Molin D, Bigot M, et al. DMGD-Compensated Links//Optical Fiber Communication Conference, OSA Technical Digest (online) (Optical Society of America, 2017), paper Tu2J.4.

[19] Li G F, Bai N, Zhao N B, et al. Space-division multiplexing: the next frontier in optical communication. Advances in Optics and Photonics, 2014, 6(4): 413～487.

[20] Bigot-Astruc M, Boivin D, Sillard P. Design and fabrication of weakly-coupled few-modes fibers. Photonics Society Summer Topical Meeting Series, 2012:189-190 (TuC1.1).

[21] Sillard P and Molin D. A review of few-mode fibers for space-division multiplexed transmissions. European Conference & Exhibition on Optical Communication, 2013:1～3 (Mo.3.A.1).

[22] Sillard P, Bigot-Astruc M, Molin D. Few-mode fibers for mode-division-multiplexed systems. Journal of Lightwave Technology, 2014, 32(16): 2824～2829.

[23] Xia C, Amezcua-Correa R, Bai N, et al. Hole-assisted few-mode multicore fiber for high-density space-division multiplexing. IEEE Photonics Technology Letters, 2012, 24: 1914～1917.

[24] Ryf R, Bolle C A, von Hoyningen-Huene J. Optical coupling components for spatial multiplexing in multi-mode. European Conference and Exhibition on Optical Communication, 2011, Th.12.B.1: 1～3.

[25] Fontaine N K. Photonic lantern spatial multiplexers in space-division multiplexing. Photonics Society Summer Topical Meeting Series, 2013: 97～98.

[26] Ryf R, Fontaine N K, Montoliu M, et al. Photonic-lantern-based mode multiplexers for few-mode-fiber transmission. Optical Fiber Communication Conference, Optical Society of America, 2014.

[27] Leon-Saval S G, Fontaine N K, Salazar-Gil J R, et al. Mode-selective photonic lanterns for space-division multiplexing. Opt. Express, 2014, 22: 1036～1044.

[28] Fontaine N K, Leon-Saval S G, Ryf R, et al. Mode-selective dissimilar fiber photonic-lantern spatial multiplexers for few-mode fiber. Proceedings of European Conference on Optical Communication, 2013.

[29] Huang B, Fontaine N K, Ryf R, et al. All-fiber mode-group-selective photonic lantern using graded-index multimode fibers. Opt. Express, 2015, 23: 224～234.

[30] Huang B. Mode-group-selective photonic lantern using graded-index multimode fibers//Optical Fiber Communication Conference, OSA Technical Digest (online) (Optical Society of America, 2015), paper W2A.9.

[31] Guan B, Ercan B, Fontaine N K, et al. Mode-group-selective photonic lantern based on integrated 3D devices fabricated by ultrafast laser inscription//Optical Fiber Communication Conference, OSA Technical Digest (online) (Optical Society of America, 2015), paper W2A.16.

[32] Yao J. Microwave photonics. J. Lightwave Technol, 2009, 27: 314～335.

[33] Urick V J, Bucholtz F, McKinney J D, et al. Long-haul analog photonics. J. Lightwave Technol, 2011, 29: 1182～1205.

[34] Wen H, Zheng H J, Zhu B Y, et al. Experimental demonstration of long-distance analog transmission over few-mode fibers. OSA, OFC2015, M3E.2, 2015.3.22-26, Los Angeles, California, USA.

[35] Wen H, Zheng H J, Mo Q, et al. Analog fiber-optic links using high-order fiber modes. ECOC2015, P.7.7, 0269, Valencia (Spain), 2015.9.27～10.1.

[36] Effenberger F J. Space division multiplexing in access networks. Proc. SPIE, 2015, 9387: 938704-1～938704-6.

[37] Cen X, Chand N, Velazquez-Benitez A M, et al. Demonstration of world's first few-mode GPON. Proc.Eur. Conf. Opt. Commun, 2014: 1～3.

[38] Xia C, Chand N, Velázquez-Benítez A M, et al. Time-division-multiplexed few-mode passive optical network. Opt. Exp., 2015, 23: 1151～1158.

[39] Li B, Feng Z, Tang M, et al. Experimental demonstration of large capacity WSDM optical access network with multicore fibers and advanced modulation formats. Opt. Exp., 2015, 23(9): 10997～11006.

[40] Ren F, Li J, Hu T, et al. Cascaded mode-division-multiplexing and time-division-multiplexing passive optical network based on low mode-crosstalk FMF and mode MUX/DEMUX. IEEE Photon. J.,2015, 7(5): 1～9.

[41] Wen H, Xia C, Velazquez-Benitez A, et al. First demonstration of 6-mode PON achieving a record gain of 4 dB in upstream transmission loss budget. Journal of Lightwave Technology, 2016, 34(8): 1990～1996.

[42] Xia C, Wen H, Velázquez-Benítez A M, et al. Experimental demonstration of 5-mode PON achieving a net gain of 4 dB in upstream transmission loss budget. ECOC2015, Tu.1.5.2, 0520, Valencia (Spain), 2015.9.27～10.1.